Digital Design Techniques and Exercises

Vaibhav Taraate

Digital Design Techniques and Exercises

A Practice Book for Digital Logic Design

Springer

Vaibbhav Taraate
VLSI Design
1 Rupee S T (Semiconductor
Training @ Rs.1)
Pune, Maharashtra, India

ISBN 978-981-16-5954-6 ISBN 978-981-16-5955-3 (eBook)
https://doi.org/10.1007/978-981-16-5955-3

This Springer imprint is published by the registered company Springer Nature Singapore Pte Ltd.
The registered company address is: 152 Beach Road, #21-01/04 Gateway East, Singapore 189721, Singapore

Dedicated to my Inspiration Respected
Bharat Ratna J. R. D. Tata
and
Ratan Tata

Preface

The understanding of the digital design elements and their role in the design of the digital system is especially important to the logic designers, system design engineers, RTL design engineers and even to the ASIC/FPGA design engineers.

There are many evolutions in the digital design during the past century, and the main objective of this book is to have a discussion on the important design techniques useful in the digital design.

The book has 12 chapters and is mainly useful to understand about the digital design techniques, FSM-based designs, design optimization, data and control path designs, timing of the design, area and speed requirements and optimization during logic design phase. Few of the logic diagrams and timing sequences are captured using xilinx ISE and Vivado tools. For more information about the FPGA EDA tools please visit www.xilinx.com !

The book even covers the advanced concepts like architecture design, multiple clock domain designs, multiple power domains, system design and interfacing techniques.

The book has exercises at the end of each chapter and is useful to apply the design concepts.

Chapter 1 "Basics of Digital Design" discusses about the basics of the digital design elements, Boolean function implementation techniques and main important goals during the design.

Chapter 2 "Design Using Universal Logic" the universal logic elements and their use in the design aer discussed. The chapter discusses about the cascade and parallel logic and the design techniques to improve the speed of the design and to optimize for the area.

Chapter 3 "Combinational Design Resources" discusses about the various code converters, combinational design resources and the arithmetic resources. The design techniques discussed are useful to design the combinational or glue logic. The chapter even focuses on the various performance improvement techniques and their use to design the combinational logic.

Chapter 4 "Case Study: ALU Design" discusses about the use of the combinational resources and arithmetic elements to design the digital circuits. The objective is to optimize the design to have the least area and maximum speed. The chapter discusses about the basics of the instruction processing and the optimization for the area and speed.

Chapter 5 "Practical Scenarios and the Design Techniques" discusses about the parallel versus cascade, priority logic and their use in the design. The design of the combinational logic using the decoders and encoders is also discussed in this chapter.

Chapter 6 "Basics of the Sequential Design," the sequential design elements are latches and flip-flops and they are extensively used in the design. The latch-based designs and flip-flop-based designs and their applications are discussed in this chapter.

Chapter 7 "Sequential Design Techniques" discusses about the various techniques useful to implement the sequential designs. The goal is to have the sequential design which has lesser area, maximum speed and low power. The chapter is useful to understand the sequential design techniques to design the counters, registers with the goal of area and speed optimization.

Chapter 8 "Important Design Scenarios" discusses the important design scenarios and techniques useful to design the sequential logic. The chapter is useful to understand about the duty cycle and how to design the sequential circuits with the goal to have duty cycle control.

Chapter 9 "FSM Design Techniques" discusses about the FSM design techniques and their applications in the digital design. The chapter is useful to understand the Moore and Mealy machine designs, encoding methods and their use in the design.

Chapter 10 "Advanced Design Techniques-1" discusses about the data and control path designs and the timing of the synchronous sequential circuits. Even this chapter focuses on the various advanced design techniques which are useful to optimize for the area, speed and power. These techniques we can use in the design of the architecture and also in the high-speed digital designs.

Chapter 11 "Advanced Design Techniques-2" focuses on the architecture design for the given functional specifications. The chapter is even useful to understand about the design specific scenarios like multiple clock domains, multiple power domains, synchronizers, design specific scenarios and the performance improvement for the design.

Chapter 12 "System Design and Considerations" discusses about the use of the digital design techniques in the system design, IO and memory interfacing and other important goals

The book includes many practical design scenarios and techniques. The book is useful to understand the design techniques and important design and optimization scenarios. The book even covers the performance improvement strategies and techniques at the logic and architecture level.

This book is useful to the engineering students, digital design engineers, VLSI beginners and professionals those who wish to design the digital systems!

Pune, India Vaibhav Taraate

Acknowledgements

Most of the engineers requested me to write a book on *Digital Design Techniques and Exercises* during the corporate programs. Over the period of time whatever experience which I have gained I thought to document in this manuscript.

This book is possible due to the help of many people. I am thankful to all the participants to whom I taught the subject "Digital Design in VLSI perspective" in various multinational corporations. I am thankful to all those entrepreneurs, design/verification engineers and managers with whom I worked in the past almost around 20 years.

I am thankful to my dearest friends for their constant support. Especially, I am thankful to my students, friends, well-wishers and my family members. Special thanks to Neeraj, Deepesh, Jyoti, Suman and Annu for their best wishes and valuable help during the manuscript work.

Special thanks to Somi, Siddhesh and Kajal for their faith and belief on me and for their better support during manuscript work. Especially thankful to Ravi, Divya, Swati, Rahul for their best wishes !

Finally, I am thankful to Springer Nature staff, especially Swati Meherishi, Muskan Jaiswal, Ashok Kumar, Silky Sinha for their great support during the various phases of the manuscript.

Special thanks in advance to all the readers and engineers for buying, reading and enjoying this book!

Acknowledgements

Contents

About the Author

Vaibbhav Taraate is Entrepreneur and Mentor at "1 Rupee S T". He holds a BE (Electronics) degree from Shivaji University, Kolhapur in 1995 and secured a gold medal for standing first in all engineering branches. He has completed his M.Tech. (Aerospace Control and Guidance) in 1999 from Indian Institute of Technology (IIT) Bombay. He has over 18 years of experience in semi-custom ASIC and FPGA design, primarily using HDL languages such as Verilog and VHDL. He has worked with few multinational corporations as consultant, senior design engineer, and technical manager. His areas of expertise include RTL design using VHDL, RTL design using Verilog, complex FPGA-based design, low power design, synthesis/optimization, static timing analysis, system design using microprocessors, high speed VLSI designs, and architecture design of complex SOCs.

Chapter 1
Basics of Digital Design

The basics of digital logic design and the various efficient techniques are useful during the optimization of the combinational logic.

Most of the time, we experience the need of the digital design techniques to design the digital systems. If we consider any digital system, then the understanding of the digital design techniques, and their use is helpful to the engineers to design and implement the systems. The main considerations are the area, speed and power requirements for these systems and their efficient understanding while implementing the digital systems. In this context, the chapter discusses the basics of the digital design techniques with their main important goals.

1.1 Digital Logic and the Evolution

The digital logic has evolved during last century and has various efficient techniques. Most of the time, we have used the techniques for the Boolean equation simplifications and for the optimization. The main important techniques are very basic, and they are

1. Boolean theorems
2. Sum of Product (SOP) and Product of Sum (POS) simplifications
3. Karnaugh map (K-map)
4. De Morgan's theorems
5. Design optimization techniques
6. Delay optimization techniques
7. Power optimization techniques
8. Speed optimization techniques.

Most of us are familiar with these techniques, and we use these during the various design stages such as architecture and micro-architecture of the design.

V. Taraate, *Digital Design Techniques and Exercises*,
https://doi.org/10.1007/978-981-16-5955-3_1

If we consider the decade of 80s, then we have witnessed migration of EDA tool flow from the schematic entry to the hardware description language (HDL) during 1984–1985. Most of the EDA tool companies evolved the algorithms using the Verilog and VHDL languages to carry out the design and implementation.

Even we have witnessed few PLD-based designs and use of FPGA tools during that decade. In such scenarios, the book is useful to understand various design techniques from basics to complex designs. How to sketch the design architecture and micro-architecture for the design and how to use the advanced digital design techniques is also discussed in this book.

The next few sessions are useful to understand about the basic digital design elements and their role in the design.

1.2 The Important Considerations

As most of us are familiar that, the digital design operates on the binary data, we will consider the bit as binary digit, and it has logic 0 and logic 1 values. Logic 0 stands for the VSS (GND) and logic 1 for VDD or VCC. The digital design has classified into the category as follows:

1. **Combinational logic**: In the combinational logic an output is the function of the present input. If input changes, then an output will change after the propagation delay of the combinational logic, hence avoiding the cascading of stages! Following are examples of the combinational logic:

   a. **Logic gates**
   b. **Arithmetic resources**
   c. **Multiplexers**
   d. **Decoders**
   e. **Demultiplexers**
   f. **Encoders**

2. **Sequential logic**: In the sequential design an output is the function of the present input and past output. Examples are latches and flip-flop used to design the sequential logic. The examples of the sequential logic are

   a. **Latches**
   b. **Flip-flops**
   c. **Counters**
   d. **Shift registers**
   e. **Memory elements**

While designing the digital logic, the main important considerations are **area, speed and power**. Even we need to incorporate the **concurrency, parallelism** and the **pipelining** depending on the design goals.

Most of the chapters discuss the use and application of these elements and the design techniques useful to improve the speed, power and area while using these elements!

1.2.1 *Area of the Design*

Area of any design is the number of logic gates or number of logic elements used in the design. The density of logic is how many logic gates in the unit-square area. For example, the processor area is 100,000 logic gates. The design engineers need to use the various design techniques to improve the area of design. Few of the important design techniques are discussed in this book. These techniques are mainly

1. Resource sharing
2. Logic duplications for FPGA-based designs
3. Splitting the larger combinational designs
4. Mux-based logic versus gate-based designs
5. Resource optimization at the architecture level.

1.2.2 *Speed of the Design*

The speed of the design is one of the important parameters, and the speed is limited due to the inertial gate delay or due to cascading of the number of logic stages. As shown in Fig. 1.1 if we consider the CMOS NOT gate then at the output there is formation of the stray capacitance and the inertial delay of the NOT gate is due to the time required for charging and discharging of the capacitor. The inertial delay is

Fig. 1.1 CMOS NOT gate

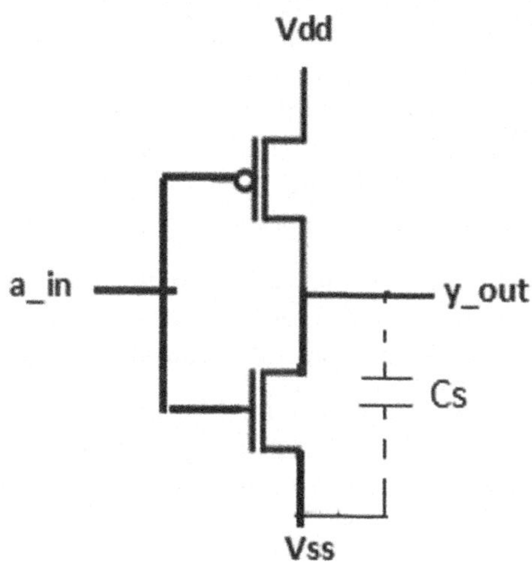

the propagation delay and defined as the amount of time required for the output to get the valid logic level after change in the input.

For the sequential design, the speed mainly depends upon the timing parameters of the sequential design elements such as setup time, hold time and clock to q delay (propagation delay of the flip-flop). Refer Chap. 10 for more details about the timing parameters.

There are various speed improvement techniques used during the design stages, and few important techniques which are discussed in this book are

1. Balancing the register timing
2. Pipelining
3. Register balancing
4. Optimizing for the timing paths in the design
5. Use of the parallelism if area is not important consideration.

1.2.3 Power

The power is an important design considerations. Most of we have come across the terms such as switching power, static and dynamic power. The power dissipation should be minimum, and the book also focuses on the low power architecture design techniques and their role at various stages of the design. Consider Fig. 1.1, as shown as the stray capacitance is formed at the output, the power is specified as

$$\text{Power} = \alpha C V^2 f$$

where α = switching factor

C = load or stray capacitance

V = supply voltage and f = frequency

To have the minimum power the load capacitance should be minimum, the voltage should be minimum, but as we cannot compromise on the speed of the design, we need to balance between the desired power and speed of the design.

There is always the trade-off between the speed and power, and the main goal of the logic design team is to have the balance act to achieve the desired speed and power for the design.

There are various power optimization techniques, and few of them are

1. Use of low power cells in the design
2. Low power aware architecture design
3. Clock gating for the sequential logic.

These techniques are discussed in the subsequent chapters.

Now as we have understood the goals and objectives during the logic design and to use the desired techniques, let us start from the basic logic elements. As stated earlier, the basic logic elements are logic gates, and the following section discusses them.

1.3 Logic Gates

The logic gates are important logic design elements. The focus of this book is more on the use of the logic elements to design the area, power and speed-efficient logic. As name indicates that the logic gates are used to perform the desired logic function. The logic gates have inputs and outputs, and they are used to build the combinational and sequential logic.

Although most of the engineers are familiar with the logic gates, let us discuss them in the context of the logic design and optimization of the logic!

1. **NOT Gate**

NOT is complement of the input and also called as logic inverter. It has single input and single output. It just complements the binary input.

The truth table of the NOT gate is shown in Table 1.1. It has input a_in and an output y_out. The relationship between the input and output is given by

$$y\_out = \overline{a\_in}$$

The symbolic representation is shown in Fig. 1.2, and the output is complement of an input. The logic 1 complement is logic 0 and vice versa.

2. **OR Gate**

OR is the logical OR of the inputs and in simple words indicates that this or this! The two input OR gate performs logical OR on the two binary inputs to generate a single bit binary output.

The truth table of the OR gate is shown in Table 1.2. It has inputs a_in, b-in and an output y_out. The relationship between the inputs and output is given by

$$y\_out = a\_in + b\_in$$

Figure 1.3 is symbolical representation of OR gate, and it indicates that either a_in or b_in should be logic 1 to get an output as logic 1. Hence, the logic is OR logic.

Table 1.1 Truth table of NOT gate

a_in	y_out
0	1
1	0

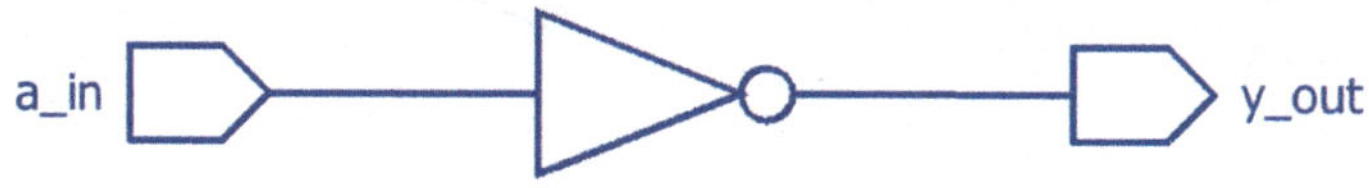

Fig. 1.2 NOT gate

Table 1.2 Truth table of OR gate

a_in	b_in	y_out
0	0	0
0	1	1
1	0	1
1	1	1

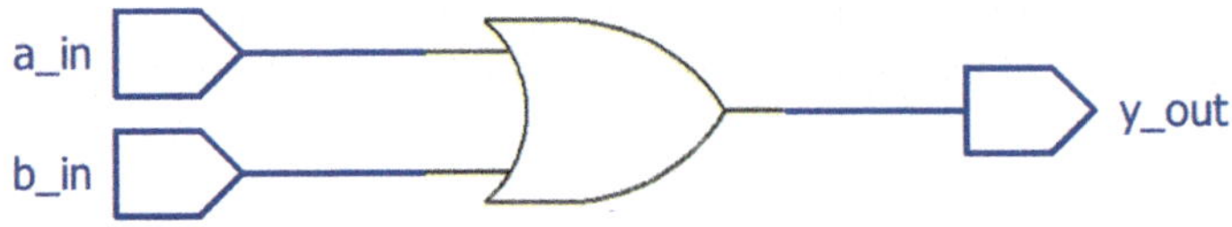

Fig. 1.3 OR gate

3. **NOR Gate**

The NOR is NOT of OR, and the output of NOR gate is logic 1 when all the inputs are logic 0. If one of the input is logic 1, then an output of NOR gate is logic 0. The NOR gate is universal gate because by using the minimum number of NOR gates, any Boolean function can be implemented.

The truth-table of NOR gate is shown in Table 1.3 and has inputs as a_in, b_in and output as y_out. The relationship between the inputs and an output is given by

$$y\_out = \overline{a\_in + b\_in}$$

The NOT of OR is shown in Fig. 1.4 which is the cascade of OR and NOT. The issue is the larger propagation delay due to cascading of the OR and NOT. So, during design, avoid the cascade logic. If we consider the delay of each logic gate is 0.5 ns, then the propagation delay of the NOR logic is 1 ns.

Table 1.3 Truth-table of NOR gate

a_in	b_in	y_out
0	0	1
0	1	0
1	0	0
1	1	0

Fig. 1.4 NOT of OR

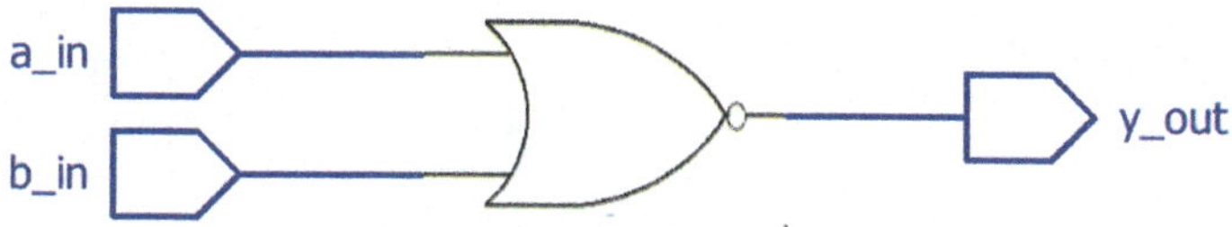

Fig. 1.5 NOR gate

The symbol of the NOR gate is shown in Fig. 1.5, and the minimum number of NOR gates should be used to implement the Boolean function. ***By using the minimum number of NOR gates, any Boolean function can be realized, and hence, the NOR gate is called as universal gate.***

4. **AND Gate**

The AND logic gate output is logic 1 when both the inputs a_in and b_in are at logic 1. Hence, the logic is represented by using a_in AND b_in. The logic expression of 2-input AND gate is given by

$$y\_out = a\_in \cdot b\_in$$

The · (pronounced as dot) indicates the AND operation. The truth-table is shown in Table 1.4. As described in the truth-table when both the inputs a_in and b_in are logic 1, an output of the AND gate is logic 1. When one of the input is logic 0, the output of AND gate remains at logic 0.

The symbol of the 2-input AND is shown in Fig. 1.6, and as shown, the AND gate has two inputs a_in, b_in and an output as y_out.

5. **NAND Gate**

NAND is NOT of AND, and the output of NAND is logic 0 when both the inputs are at logic 1. If one of the input of NAND gate is at logic 0, then an output of NAND

Table 1.4 Truth table of AND gate

a_in	b_in	y_out
0	0	0
0	1	0
1	0	0
1	1	1

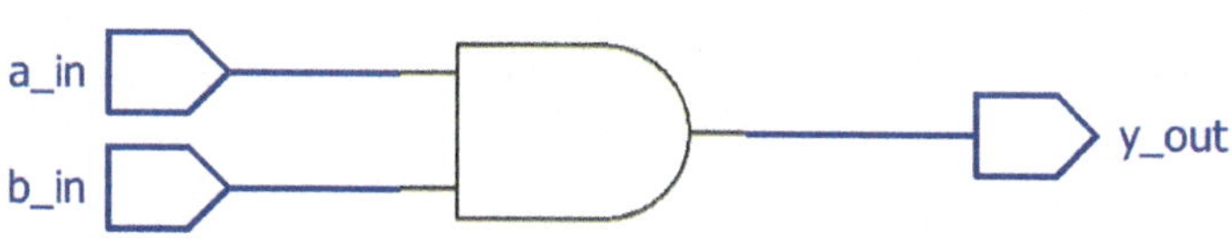

Fig. 1.6 AND gate

Table 1.5 Truth table of NAND gate

a_in	b_in	y_out
0	0	1
0	1	1
1	0	1
1	1	0

Fig. 1.7 NOT of AND

gate is logic 1. The truth-table of 2-input NAND gate is described in Table 1.5. The logic expression is given by

$$y\_out = \overline{a\_in \cdot b\_in}$$

The cascade of the AND, NOT is shown in Fig. 1.7. As discussed earlier, the design engineers should avoid the cascading of the stages.

The NAND gate symbolical representation is shown in Fig. 1.8, and as shown, it has two inputs a_in, b_in and an output as y_out.

6. **XOR Gate**

The XOR gate is also called as exclusive OR. The truth-table of the XOR gate is shown in Table 1.6. As shown, the output of the 2-input XOR gate is logic 1 when both the inputs are not equal. The logic expression of XOR gate is

$$y\_out = a\_in \oplus b\_in$$

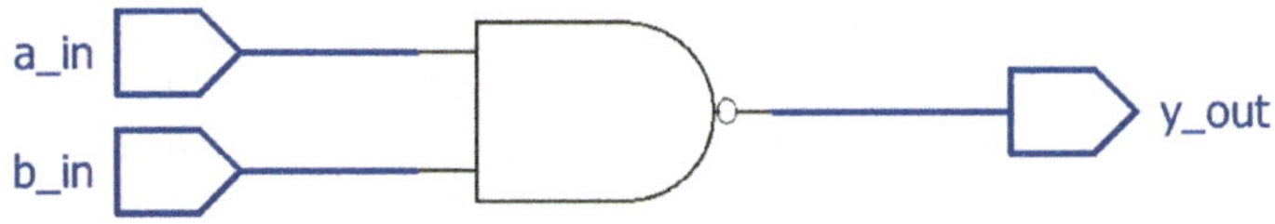

Fig. 1.8 NAND gate

Table 1.6 Truth table of XOR gate

a_in	b_in	y_out
0	0	0
0	1	1
1	0	1
1	1	0

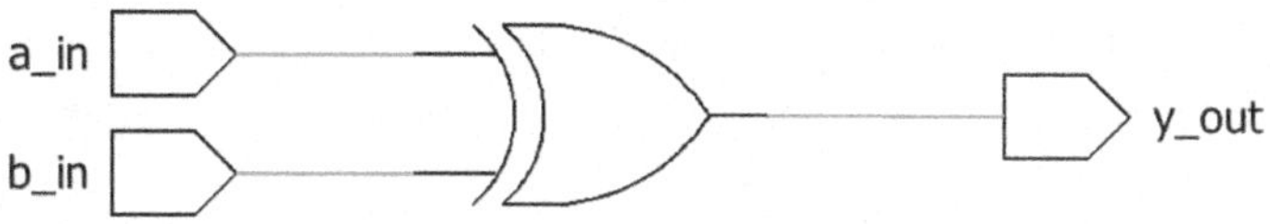

Fig. 1.9 XOR gate

The symbol of the XOR gate is shown in Fig. 1.9, and as shown, the 2-input XOR gate has inputs a_in, b_in and an output as y_out.

7. **XNOR Gate**

The XNOR is NOT of XOR. The symbol of the XNOR is ⊙ *(pronounced as EXNOR).* The logic expression is given by

$$y\_out = a\_in \odot b\_in$$

The EXNOR or exclusive NOR or XNOR these are few names which we use while implementing the design. The XNOR is cascade of the XOR and NOT and hence called as NOT of XOR. The XNOR using XOR is shown in Fig. 1.10.

The truth-table of the XNOR gate is described in Table 1.7. As described, the output of 2-input XNOR gate is logic 1 when both the inputs are equal.

As discussed earlier, avoid the cascading of the stages as it increases the propagation delay.

The symbol of the XNOR gate is shown in Fig. 1.11, and as shown, the logic has a_in, b_in inputs and output as y_out.

Fig. 1.10 NOT of XOR

Table 1.7 Truth table of XNOR gate

a_in	b_in	y_out
0	0	1
0	1	0
1	0	0
1	1	1

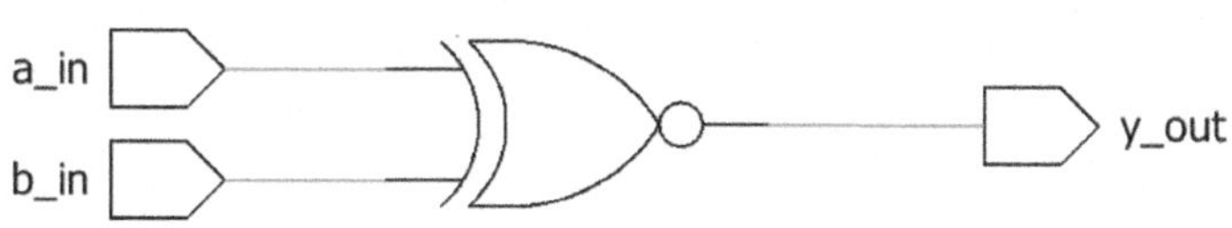

Fig. 1.11 XNOR gate

Table 1.8 Truth table of bubbled OR is equal to NAND

a_in	$\overline{\text{a\_in}}$	b_in	$\overline{\text{b\_in}}$	y_out
0	1	0	1	1
0	1	1	0	1
1	0	0	1	1
1	0	1	0	0

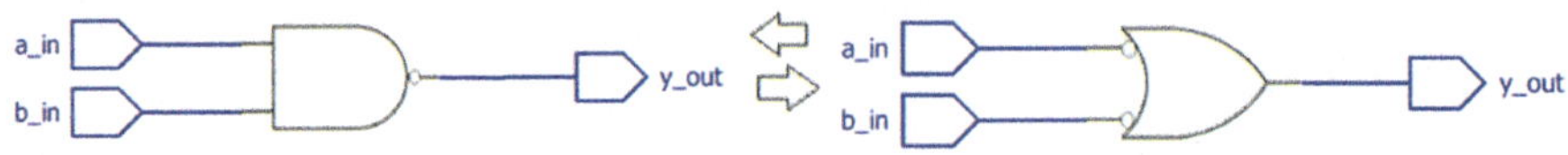

Fig. 1.12 Bubbled OR is equal to NAND

1.4 De Morgan's Theorems

As we have good understanding of the logic gates now, let us try to understand the important theorems as De Morgan's theorems. For the Boolean simplification, the two important theorems are

1.4.1 NAND is Equal to Bubbled OR

$$\overline{\text{a\_in} \cdot \text{b\_in}} = \overline{\text{a\_in}} + \overline{\text{b\_in}}$$

The truth-table is shown in Table 1.8. The output if bubbled OR matches with the output of 2-input NAND, and hence in simple words, we can consider bubbled OR is equal to NAND or vice versa.

For Boolean simplifications, we can use the De Morgan's theorems. We can use the bubbled OR as a NAND during Boolean simplifications (Fig. 1.12).

1.4.2 NOR is Equal to Bubbled AND

$$\overline{\text{a\_in} + \text{b\_in}} = \overline{\text{a\_in}} \cdot \overline{\text{b\_in}}$$

Table 1.9 Truth table of bubbled AND is equal to NOR

a_in	$\overline{\text{a\_in}}$	b_in	$\overline{\text{b\_in}}$	y_out
0	1	0	1	1
0	1	1	0	0
1	0	0	1	0
1	0	1	0	0

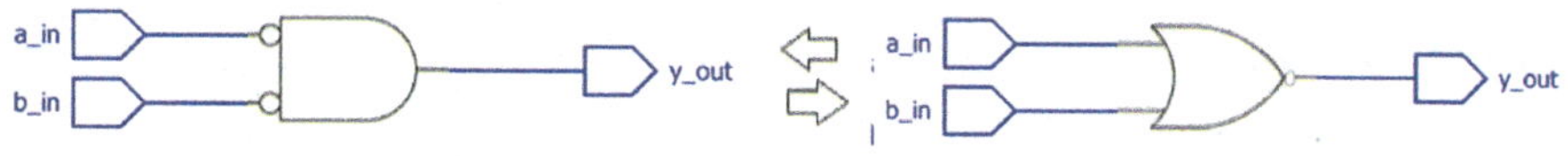

Fig. 1.13 Bubbled AND is equal to NOR

The truth-table is shown in Table 1.9. The output if bubbled AND matches with the output of 2-input NOR, and hence in simple words, we can consider bubbled AND is equal to NOR or vice versa.

For Boolean simplifications, we can use the De Morgan's theorems. We can use the bubbled AND as a NOR during Boolean simplifications (Fig. 1.13).

1.5 Multiplexer as Universal Logic

The multiplexer is many to one switch, and they are used in the multiplexing of the buses, clock multiplexing and other various applications. The multiplexer has many inputs and single output. The logic level on the select inputs decides which input is selected, and according to that, the multiplexer output is generated. The multiplexer is also called as mux, and throughout this book, we will call multiplexer as mux. The truth-table of 2:1 mux is shown in Table 1.10. The 2:1 mux has two inputs a_in, b_in and single select input sel_in.

As described, the output of mux is b_in for sel_in = 0, for the sel_in = 1 output is a_in.

The symbol of the 2:1 mux is shown in Fig. 1.14, and as shown depending on the sel_in status, it passes one of the input either a_in, b_in to an output y_out.

Now, let us implement the 2:1 mux using the logic gates. From the truth table, we can get the product term, and we can use the Sum of Product (SOP) expression to implement the 2:1 mux (Fig. 1.15).

Table 1.10 Truth table of 2:1 MUX

sel_in	y_out
0	b_in
1	a_in

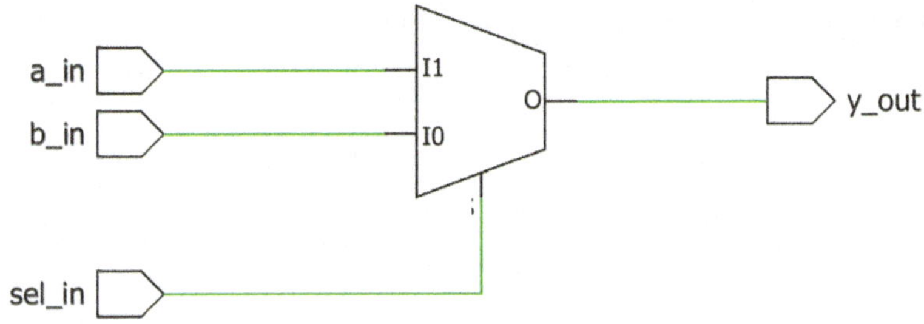

Fig. 1.14 Symbolical representation of 2:1 MUX

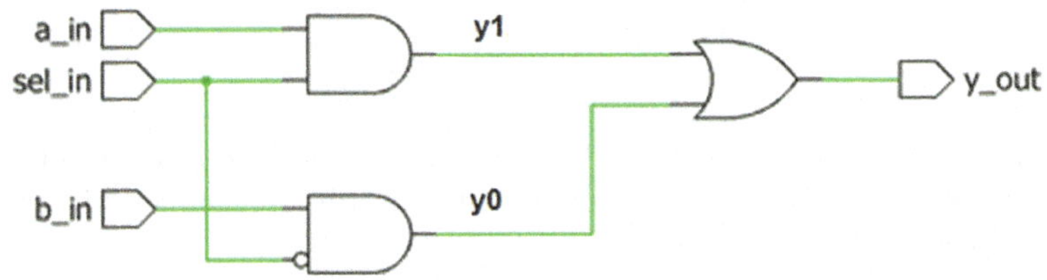

Fig. 1.15 Gate-level structure of 2:1 MUX

sel_in	y_out	Product term
0	b_in	$y0 = \overline{sel\_in} \cdot b\_in$
1	a_in	$y1 = sel\_in \cdot a\_in$

Using the product term, the SOP expression for the 2:1 mux is given by

$$y\_out = y0 + y1$$

$$y\_out = \overline{sel\_in} \cdot b\_in + sel\_in \cdot a\_in$$

1.6 Optimization Goals and Applications in VLSI Context

While using the digital elements to design the digital logic, following are few of the optimization goals.

1. Do not use the cascading of the logic as it adds significant delays.
2. Use the minimum number of universal logic gates to implement the Boolean function.
3. Use the multiplexers to implement the Boolean functions.
4. Use the low power design cells or gates to improve the power.
5. Use the logic cells which has least propagation delay

In the VLSI design context, the following are goals of the designers:

1. Understand the compatibility and logic levels of gates.
2. Use the minimum number of gates to implement the Boolean function.
3. Use the parallel logic if area is not an important parameter.
4. Use the low power and high-speed cells during the design.

1.7 Exercises

Now let us use the basic fundamentals of the logic gates and let us complete following few exercises.

1.7.1 Exercise 1: Use of the Logical Expressions to Get the Logic Equivalent

For the given expression, what is the equivalent logic gate expression at output y?

Solution: *Given* $y = A + \overline{A} \cdot B$

$$y = (A + \overline{A}) \cdot (A + B)$$

$$y = (A + B)$$

1.7.2 Exercise 2: Cascade Logic and How to Get Logic Expression?

For the given cascade logic , what is the equivalent logic gate expression at output y? (Fig. 1.16).

Solution: Output of first XOR gate

$$y1 = X \oplus X = 0$$

Output of second XOR gate

$$y2 = X \oplus 0 = X$$

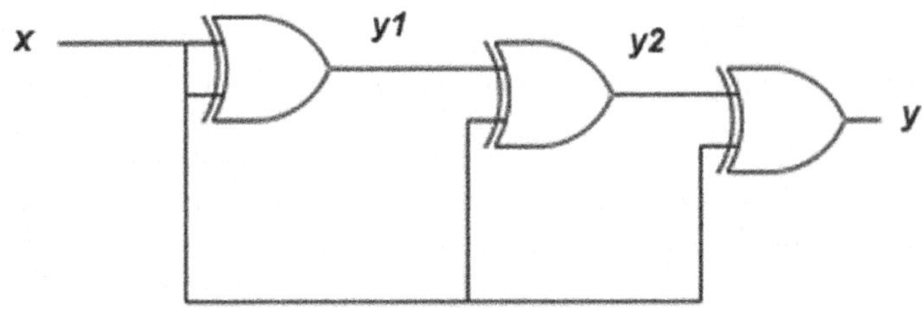

Fig. 1.16 Cascade XOR gates

Output of third XOR gate

$$y = X \oplus X = 0$$

1.7.3 Exercise 3: Complement Logic

For the given logic gate, what is the equivalent logic expression at output y? (Fig. 1.17).

Solution:

$$y = \overline{A \oplus 0}$$

$$y = \overline{A} \cdot \overline{0} + A \cdot 0$$

$$y = \overline{A} + 0$$

$$y = \overline{A}$$

Fig. 1.17 XNOR gate

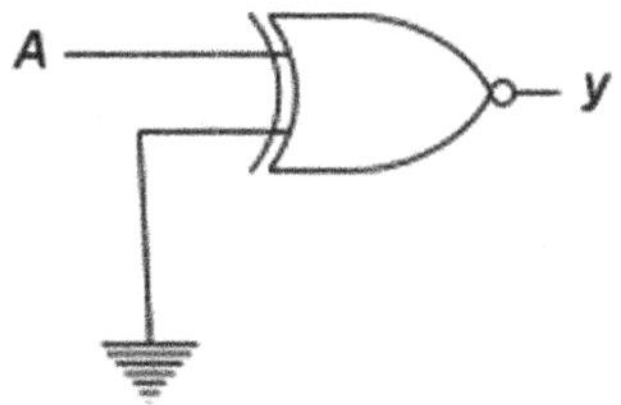

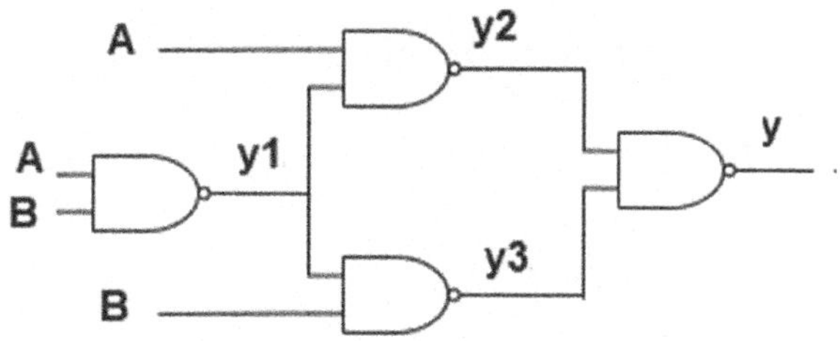

Fig. 1.18 Logic realization using NAND

1.7.4 Exercise 4: Logic Expression for the Cascade Logic

For the given cascade logic , what is the equivalent logic expression at output y? (Fig. 1.18).

Solution: *Let us find the Boolean equation for y1, y2, y3*

$$y1 = \overline{AB}$$

$$y2 = \overline{A \cdot \overline{AB}} = \overline{A \cdot \left(\overline{A} + \overline{B}\right)} = \overline{A \cdot \overline{B}}$$

$$y3 = \overline{B \cdot \overline{AB}} = \overline{B \cdot \left(\overline{A} + \overline{B}\right)} = \overline{B \cdot \overline{A}}$$

$$y = \overline{y2 \cdot y3} = \overline{\overline{(A \cdot \overline{B})} \cdot \overline{(B \cdot \overline{A})}}$$

Using De Morgan's theorem that is NAND is equal to bubbled OR, we will get $y = \overline{\overline{(A \cdot \overline{B})} \cdot \overline{(B \cdot \overline{A})}}$

$$y = \overline{\overline{(A \cdot \overline{B})}} + \overline{\overline{(B \cdot \overline{A})}}$$

$$y = A \cdot \overline{B} + \overline{A} \cdot B$$

1.7.5 Exercise 5: Output Expression for the Cascade Logic

Find the output expression for the cascade logic shown in Fig. 1.19. Consider even number of XOR gates cascaded as shown.

Solution: Output of first XOR gate $= \overline{X}$

Output of second XOR gate $= \overline{X} \oplus X$

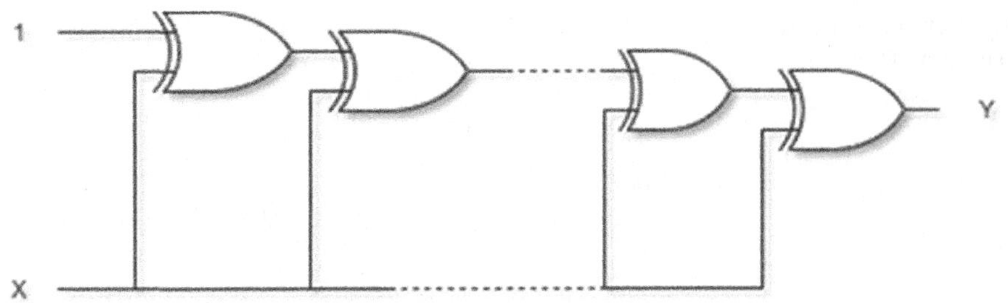

Fig. 1.19 Even number of cascade XOR gates

$$= \overline{X} \cdot \overline{X} \oplus X \cdot X = 1$$

As an even number of XOR gates are connected in cascade, the output of second XOR gate and the last XOR gate are same which is logic 1.

1.7.6 Exercise 6: Propagation Delay for the Cascade Logic

Consider the propagation delay of each XOR gate is 0.5 ns and the ten XOR gates are connected in cascade. The propagation delay at the output y of the logic is of how many ns? (Fig. 1.20).

Solution: The number of cascade stages are $n = 10$;each stage has the propagation delay (tpd) of 0.5 ns.

So, the cascade logic has propagation delay of n * tpd = 10 * 0.5 ns = 5 ns.

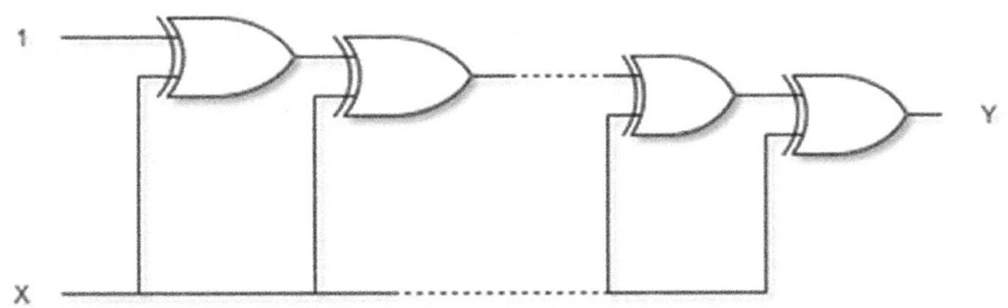

Fig. 1.20 Even gates in cascade

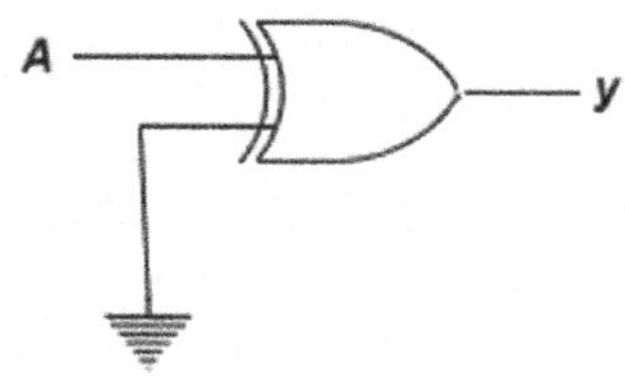

Fig. 1.21 XOR gate

1.7.7 Exercise 7: Logic Gate Output Expression

For the logic gates shown, what is the equivalent logic expression at output y? (Fig. 1.21).

Solution:

$$y = A \oplus 0$$

$$y = \overline{A} \cdot 0 + A \cdot \overline{0}$$

$$y = 0 + A$$

$$y = A$$

1.7.8 Exercise 8: Propagation Delay for the Cascade Logic

For the following logic, if each gate has propagation delay of 1 ns the maximum propagation delay to get output y is equal to? (Fig. 1.22).

Solution: Let us divide the logic shown into three regions as shown in Fig. 1.23.

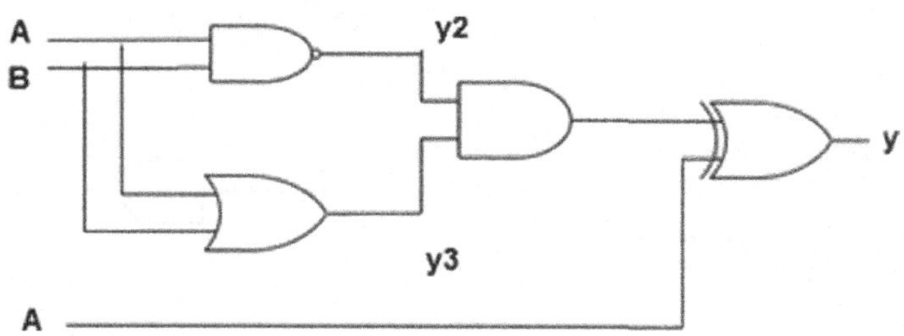

Fig. 1.22 Logic schematic

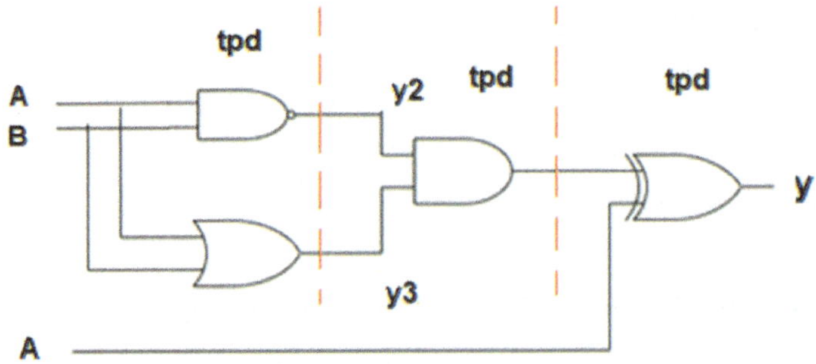

Fig. 1.23 Logic regions and delays

As the NAND and OR gate are in parallel the delay is tpd, the delay of AND gate is tpd and XOR gate is tpd, as the gate delay of each gate is 1 ns.

The overall propagation delay to get output y is 3 * tpd = 3 * 1 ns = 3 ns.Y

1.7.9 Exercise 9: The Equivalent Logic Expression

For Fig. 1.24, the equivalent logic gate expression at output y_out is?

Solution: Use the De Morgan's theorem and find the y_out.

As we know that bubbled OR is NAND

$$\overline{\text{a\_in} \cdot \text{b\_in}} = \overline{\text{a\_in}} + \overline{\text{b\_in}}$$

NOT of NAND is AND

$$\text{y\_out} = \overline{\overline{\text{a\_in} \cdot \text{b\_in}}}$$

$$\text{y\_out} = \text{a\_in} \cdot \text{b\_in}$$

So, the above figure logic gate expression is AND logic gate.

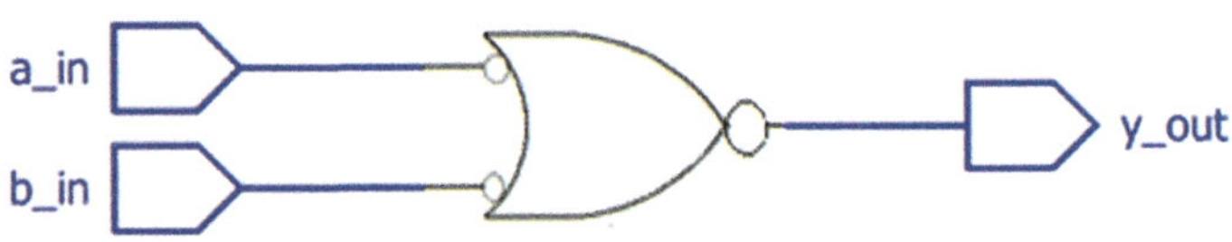

Fig. 1.24 Logic gate-1

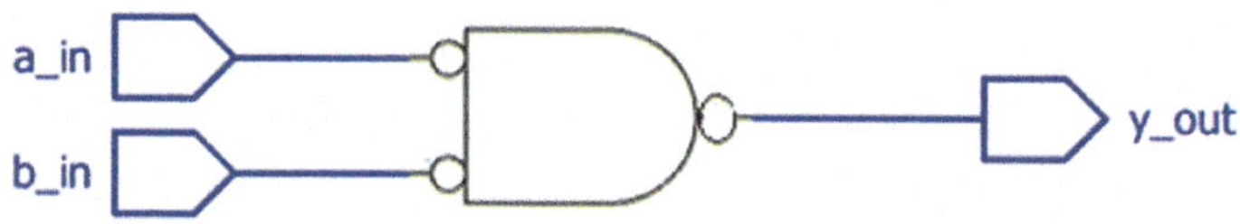

Fig. 1.25 Logic gate-2

1.7.10 Exercise 10: The Equivalent Logic Gate

For Fig. 1.25, the equivalent logic expression at output y_out is?

Solution: Use the De Morgan's theorem and find the y_out.

As we know that bubbled AND is NOR

$$\overline{\text{a\_in} + \text{b\_in}} = \overline{\text{a\_in}} \cdot \overline{\text{b\_in}}$$

NOT of NOR is OR

$$\text{y\_out} = \overline{\overline{\text{a\_in} + \text{b\_in}}}$$

$$\text{y\_out} = \text{a\_in} + \text{b\_in}$$

So, the above figure logic gate is OR gate.

1.8 Important Takeaways

Following are few of the important points to conclude this chapter.

1. The NAND and NOR are universal logic gates and the goal is to use the minimum number of logic gates to implement the Boolean functions
2. NAND logic is equivalent to bubbled OR logic.
3. NOR logic is equivalent to the bubbled AND logic.
4. Multiplexers are treated as universal logic and used to implement any kind of Boolean function.
5. The goal of the system design engineer is to implement the digital system with lesser area, lesser power and more speed.
6. Do not use the cascading stages in the design.

Chapter 2
Design Using Universal Logic

The design using universal gates and use of multiplexers as universal logic is useful during the combinational design.

The universal logic gates such as NAND, NOR, MUX and other application-specific or custom gates can be used in the design with the goal of the area optimization. The chapter is useful to understand about the role of these universal logic gates and design by using them! The chapter discusses the cascade and parallel logic and the design techniques to improve the speed of the design and to optimize for the area.

2.1 What Is Universal Logic?

Most of us are familiar with the universal logic gates. Mainly, these gates are NAND and NOR. As discussed in the previous chapter, the NAND is bubbled OR and NOR is bubbled AND. The goal of the logic design engineer is to use the minimum number of these gates and to implement the combinational logic which has minimum area. You may feel that it is difficult to directly arrive to the minimum number of the universal gates. But we can use the digital design techniques to get the minimum number!

Apart from the NAND and NOR gates, the other universal logic elements are 2:1 mux and the application-specific or custom gates. Figure 2.1 gives information about these gates, and they can be used to implement any combinational logic.

2.2 Universal Gates

Most of us are familiar with the universal logic gates as NAND and NOR. These gates are used to design the Boolean functions. The objective of the designer is to use the minimum number of logic gates. The realized logic should have the minimum

V. Taraate, *Digital Design Techniques and Exercises*,
https://doi.org/10.1007/978-981-16-5955-3_2

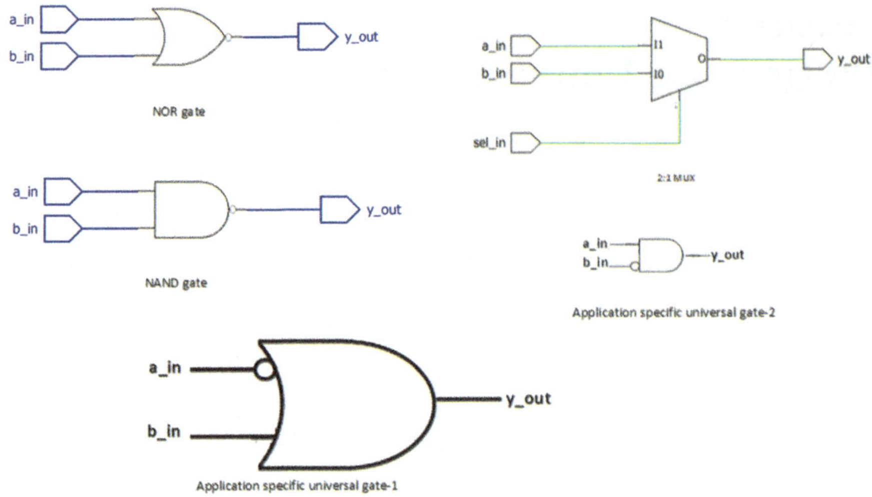

Fig. 2.1 Universal gates and mux as a universal logic

area, maximum speed and less power. The section discusses these universal logic gates and design using them.

2.2.1 NAND

As discussed in the previous chapter, NAND is NOT of AND. NAND is universal logic gate, and we can use minimum number of NAND gates to implement the Boolean function. Consider the 2-input XOR gate we can use only four NAND gates and can implement the 2-input XOR gate. The relationship between the inputs a_in, b_in of NAND gates and an output y_out is shown in Table 2.1.

Let us understand the switch-level design of the 2-input NAND gate. As we know that in the CMOS switch-level design we can use PMOS and NMOS. The PMOS passes strong 1 and used in the Vdd section. The NMOS passes the strong 0 and used in the Vss section.

As NAND is NOT of AND, in the lower section we can use the series NMOS switches and as the complement of the NMOS is PMOS, complement of the series

Table 2.1 Truth-table of 2-input NAND gate

a_in	b_in	$y\_out = \overline{a\_in \cdot b\_in}$
0	0	1
0	1	1
1	0	1
1	1	0

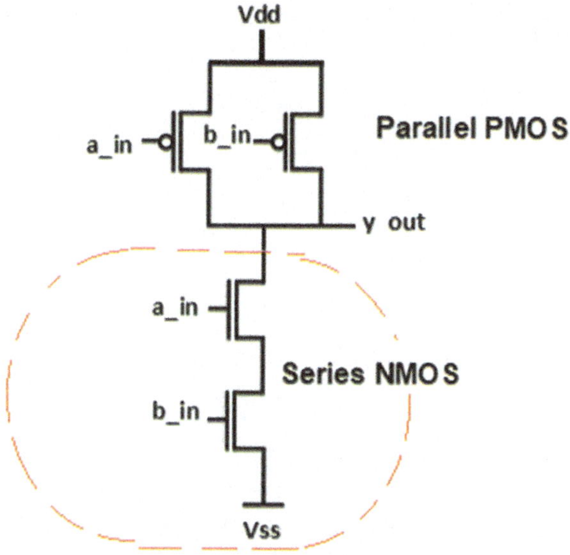

Fig. 2.2 CMOS 2-input NAND

is parallel, in the upper section we can use the parallel connection of the PMOS switches. The upper parallel section uses the supply as Vdd, and the lower series section uses the Vss. Vdd and Vss are complement of each other. The CMOS 2-input NAND gate is shown in Fig. 2.2.

2.2.2 NOR

As discussed in the previous chapter, NOR is NOT of OR. NOR is universal logic gate, and we can use minimum number of NOR gates to implement the Boolean function. Consider the 2-input XNOR gate we can use four NOR gates and can implement the 2-input XNOR gate. The relationship between the inputs a_in, b_in and an output y_out is shown in Table 2.2.

Let us understand the switch-level design of the 2-input NOR gate. As NOR is NOT of OR, in the lower section we can use the parallel NMOS switches and as the complement of the NMOS is PMOS, complement of the parallel is series, we can

Table 2.2 Truth-table of 2-input NOR gate

a_in	b_in	y_out = $\overline{\text{a\_in} + \text{b\_in}}$
0	0	1
0	1	0
1	0	0
1	1	0

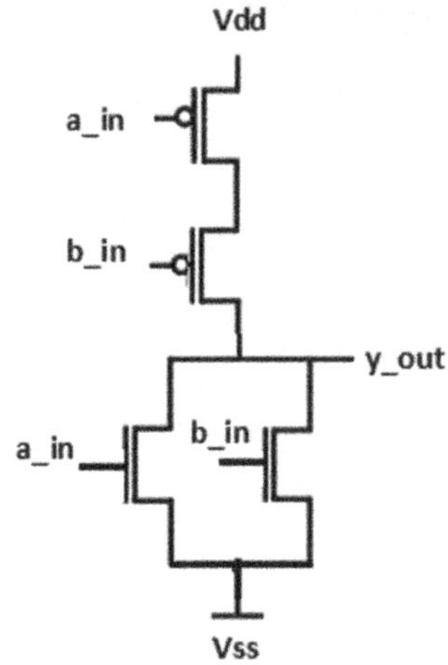

Fig. 2.3 CMOS 2-input NOR gate

use the series connection of the PMOS switches in upper section. The upper series section uses the supply as Vdd, and the lower parallel section uses the Vss. Vdd and Vss are complement of each other. The CMOS 2-input NOR gate is shown in Fig. 2.3.

2.2.3 *Other Application-Specific Universal Gates*

Most of us are familiar with the NAND and NOR as universal gates. We have come across these gates and have habit of using the minimum number of NAND or NOR gates to realize the Boolean function. As discussed in the previous chapter, we can use the minimum number of 2:1 mux to implement the boolean function. The multiplexer is also treated as universal logic.

Apart from the NAND, NOR and 2:1 mux, we can have the other application-specific gates as shown in Figs. 2.4 and 2.6 as the universal logic gates. Thease gates are also used to design the logic function.

Considering the logic gate shown in Fig. 2.4 if we wish to implement the OR gate using only the minimum number of these types of gates, then we can use following strategy (Fig. 2.5).

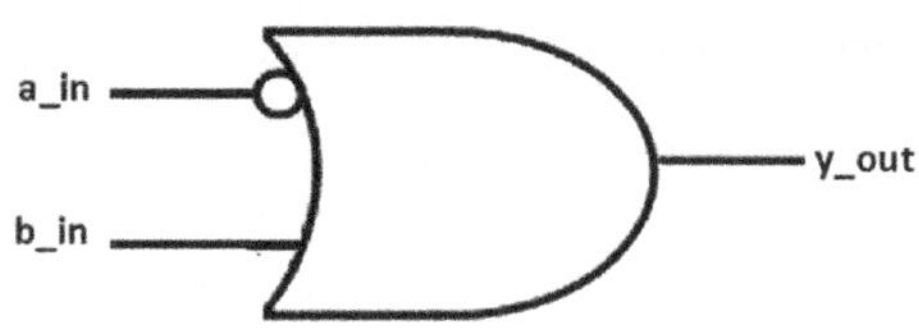

Fig. 2.4 Application-specific universal gate-1

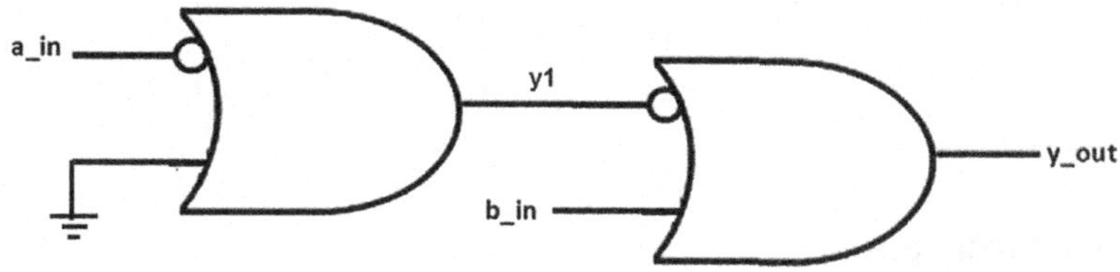

Fig. 2.5 2-input OR using the application-specific gates

Fig. 2.6 Application-specific universal gate-2

We can use two application-specific gates in cascade. Let us get the y_out expression for Fig.2.5

$$y1 = 0 + \overline{a\_in}$$

$$y1 = \overline{a\_in}$$

$$y\_out = \overline{\overline{a\_in}} + b\_in$$

$$y\_out = a\_in + b\_in$$

Consider the logic gate shown in Fig. 2.6 if we wish to implement the AND gate using only the minimum number of these types of gates, then we can use following strategy.

We can use two application-specific gates in cascade (Fig. 2.7).

$$y1 = 1 \cdot \overline{b\_in}$$

$$y1 = \overline{b\_in}$$

$$y\_out = \overline{\overline{b\_in}} \cdot a\_in$$

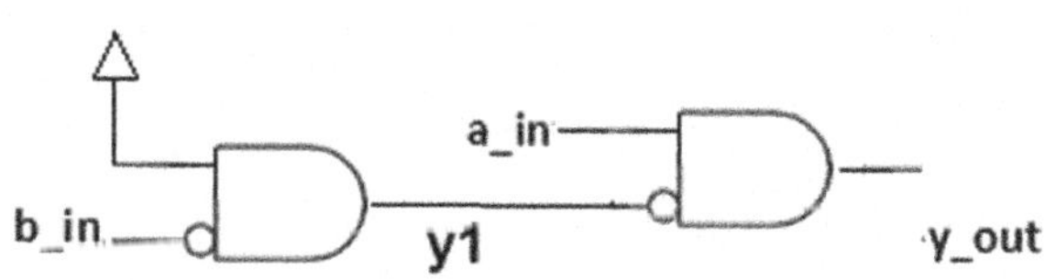

Fig. 2.7 2-input AND using the application-specific gates

$$y\_out = a\_in \cdot b\_in$$

2.3 Multiplexers

As discussed in the previous chapter, the multiplexers are treated as universal logic. By using the minimum number of multiplexers, any Boolean function can be implemented. The objective of the designer is to have least propagation delay and use of the minimum number of multiplexers. This section is useful to understand the design using minimum number of multiplexers.

2.3.1 Design Using 2:1 Mux

The design using the minimum number of 2:1 mux is discussed in this section. The 2:1 mux is the lowest input multiplexer in the hierarchy. As discussed in the previous chapter, the 2:1 mux has 2-inputs, single select line and single output line. The relationship between the input lines (m) and select lines (n) is given by $m = 2^n$.

2.3.1.1 NOT Gate Using 2:1 Mux

Now let us design the NOT gate using 2:1 mux. As shown in the table, the NOT gate has single input and single output.

Now let us compare the truth-table of NOT gate with 2:1 mux.

From comparison of Tables 2.3 and 2.4, it is clear that **sel_in = a_in, the I0 input should be connected to Vdd (logic 1) and I1 input should be connected to Vss (logic 0)**. The NOT gate using the single 2:1 mux is shown in Fig. 2.8.

Table 2.3 Truth-table of NOT gate

a_in	$y\_out = \overline{a\_in}$
0	1
1	0

Table 2.4 Truth-table of 2:1 mux

sel_in	y_out
0	I0
1	I1

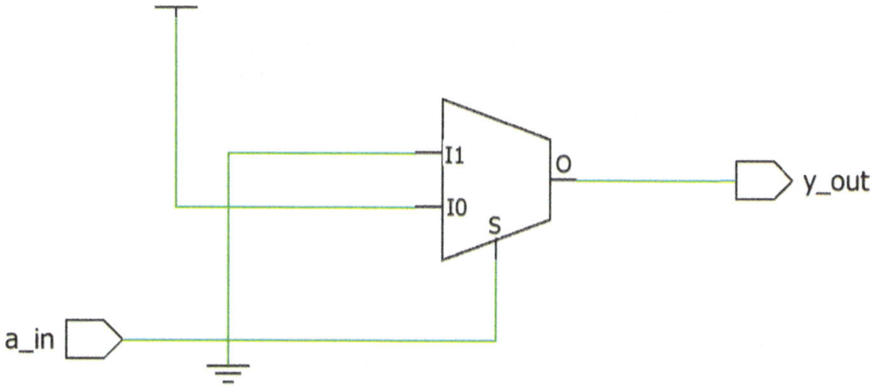

Fig. 2.8 NOT gate using 2:1 MUX

2.3.1.2 OR Gate Using 2:1 Mux

Let us design the OR gate using minimum number of 2:1 mux. Let us think, how many 2:1 mux needed to implement the OR gate? The answer is not very straight forward, and we can arrive to the implementation of OR using 2:1 mux by just observing and rearranging the truth-table of OR. So let us try to do that! (Table 2.5).

Now the better strategy is to divide the table into group of two entries as a_in is logic 0 for first two entries and logic 1 for next two entries. Now compare the b_in entries with the y_out output of OR gate. Using this strategy, let us document the entries so that we can get the equivalent truth table (Table 2.6).

The 2-input OR implementation using the single 2:1 mux is shown in Fig. 2.9. As shown when a_in which is select line of 2:1 mux is logic 1, an output of 2:1 mux is logic 1. For select line a_in = 0, an output is b_in.

Table 2.5 OR gate truth-table

a_in	b_in	y_out
0	0	0
0	1	1
1	0	1
1	1	1

Table 2.6 2-input OR using the single 2:1 mux

sel_in = a_in	y_out
0	b_in
1	1

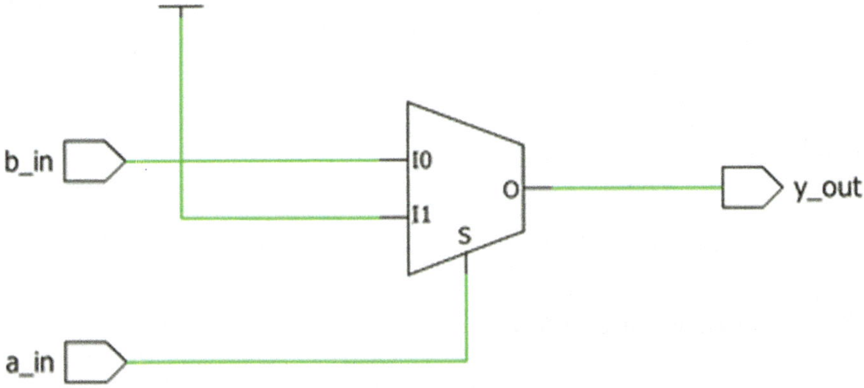

Fig. 2.9 OR gate using 2:1 MUX

2.3.1.3 NAND Using Mux

Now let us use the strategy explained in the above section to implement the 2-input NAND using minimum number of 2:1 multiplexers. Table 2.7 has four entries, and

Table 2.7 Truth-table of 2-input NAND gate

a_in	b_in	y_out
0	0	1
0	1	1
1	0	1
1	1	0

these are divided into two groups. For first two entries that is a_in = 0 if we compare b_in with y_out, then we are getting an output y_out = 1. For the next two entries if we compare b_in with y_out of NAND, then we get y_out = $\overline{\text{b\_in}}$.

Now most of the time, the beginners conclude that to implement the 2-input NAND gate, we need to have single 2:1 mux (Fig. 2.10) but that is not correct as the logic is not efficient. As shown in Table 2.8 for the select input a_in = 1, an output y_out = $\overline{\text{b\_in}}$. So, to implement the NOT of b_in, we need to have one more multiplexer.

The implementation of 2-input NAND using minimum number of 2:1 mux is shown in Fig. 2.11.

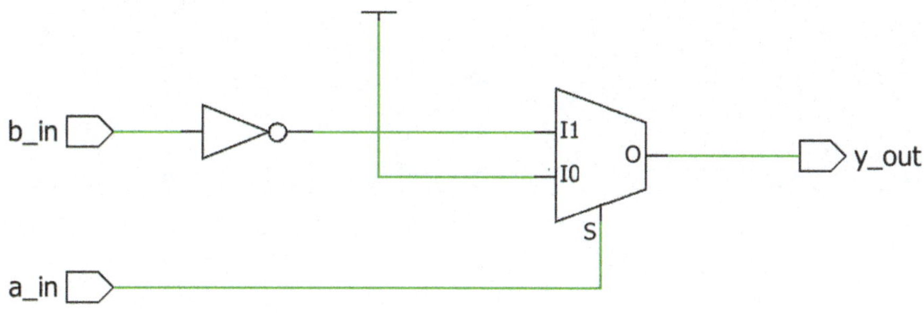

Fig. 2.10 NAND using MUX and NOT

Table 2.8 2-input NAND using multiplexer truth-table

sel_in = a_in	y_out
0	1
1	$\overline{\text{b\_in}}$

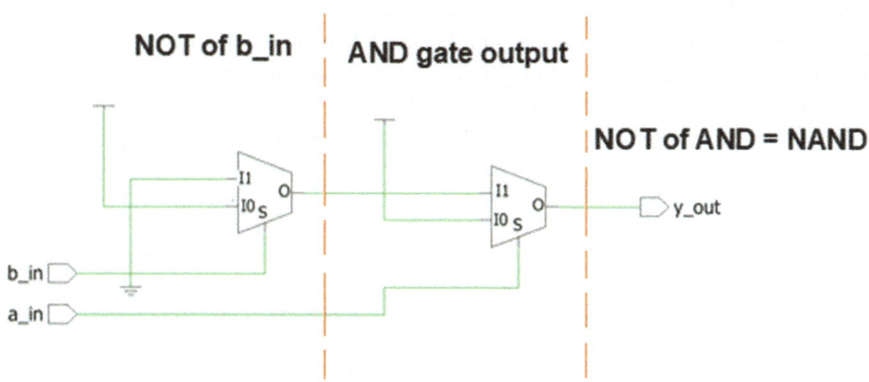

Fig. 2.11 NAND using only multiplexers

Table 2.9 Truth-table of 2-input NOR gate

a_in	b_in	y_out
0	0	1
0	1	0
1	0	0
1	1	0

2.3.1.4 NOR Using 2:1 Mux

Now let us use the strategy explained in the above section to implement the 2-input NOR using minimum number of 2:1 multiplexers. Table 2.9 has four entries, and these entries are divided into two groups. For first two entries that is a_in = 0 if we compare b_in with y_out, then we are getting an output y_out = $\overline{\text{b\_in}}$. For the next two entries if we compare b_in with y_out of NAND, then we get y_out = 0.

To implement the 2-input NOR gate, we need to have single 2:1 mux and NOT gate (Fig. 2.12) but that is not correct approach as our objective is to implement the 2-input NOR using only minimum number of 2:1 mux. As shown in Table 2.9 for the select input a_in = 0, an output y-out = $\overline{\text{b\_in}}$. So, to implement the NOT of b_in, we need to have one more 2:1 multiplexer (Table 2.10).

The implementation of 2-input NOR using only minimum number of 2:1 mux is shown in Fig. 2.13.

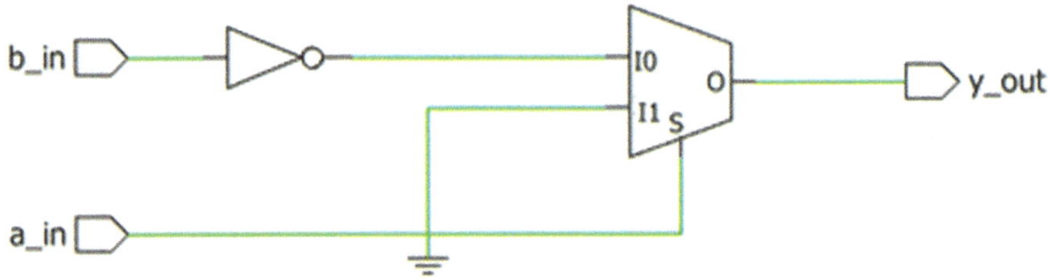

Fig. 2.12 NOR using NOT and 2:1 MUX

Table 2.10 2-input NOR table entries

sel_in = a_in	y_out
0	$\overline{\text{b\_in}}$
1	0

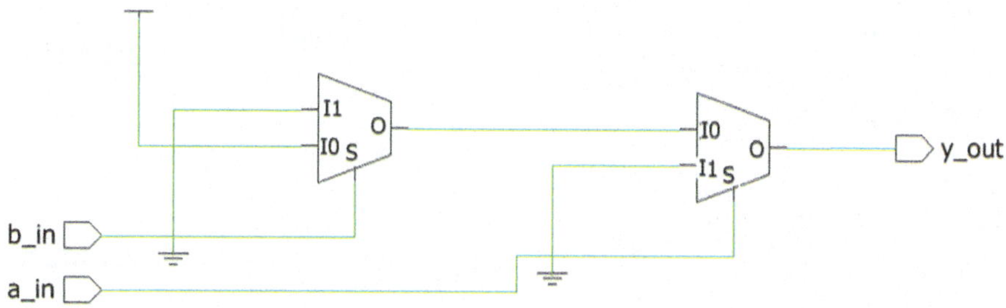

Fig. 2.13 NOR using multiplexers

Table 2.11 Truth-table of 4:1 mux

sel_in[1]	sel_in[0]	y_out
0	0	a_in
0	1	b_in
1	0	c_in
1	1	d_in

2.3.2 *4:1 MUX Using 2:1 Mux*

The 4:1 mux (Table 2.11) has two select inputs sel_in[1], sel_in[0], and depending on the status of the select inputs, one of the output a_in, b_in, c_in, d_in is connected to an output y_out. Let us use the minimum number of 2:1 mux to implement the 4:1 mux.

To realize the 4:1 mux using minimum number of 2:1 mux, let us partition the 4:1 mux table into multiple sections as shown in Table 2.12. From the partition, it is clear that we need to have three 2:1 mux to implement the 4:1 mux.

Let us now document the entries (Table 2.13) as shown below to get the output of 4:1 multiplexer (Fig. 2.14).

2.3.3 *Design Using Multiplexers*

So as discussed in the previous few sections, we can use the minimum number of 2:1 mux during the design and realization of the Boolean function. In the practical scenarios, we can use the minimum number of multiplexers to implement the

Table 2.12 Truth-table to implement the 4:1 mux using 2:1 multiplexers

sel_in[1]	sel_in[0]	y_out
0	0	a_in
0	1	b_in
1	0	c_in
1	1	d_in

2:1 MUX

2:1 MUX

2:1 MUX

Table 2.13 Output mux entries

sel_in[1]	y_out
0	y1
1	y2

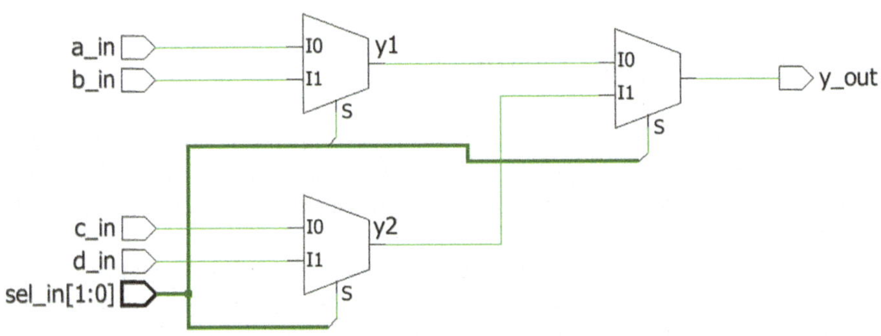

Fig. 2.14 4:1 MUX using only 2:1 multiplexers

- Boolean function that is SOP functions
- Clock muxing
- Address data bus multiplexing
- Implementation of the code converters like gray to binary and binary to gray
- Pin multiplexing

The objective of the logic designer is to use the minimum number of 2:1 mux. Even the cascade multiplexer stages reduce the speed due to increase in propagation delay.

For the better understanding and better design practice, let us discuss few exercises on the universal logic and multiplexer-based designs.

2.4 Exercises

Now let us use the understanding on the universal logic gates and the De Morgan's theorem, and let us complete the exercises with the goal of the area optimization.

2.4.1 Exercise 1: Design Using Universal Gates

Using the following custom gate, design the 2-input NAND gate (Fig. 2.15).

Solution: Let us use the understanding on the universal logic gates, and by using the minimum number of these gates, let us design the 2-input NAND gate.

As NOT of AND is NAND. Let us implement the AND and then NOT of AND (Fig. 2.16).

$$y1 = 1 \cdot \overline{b\_in}$$

$$y1 = \overline{b\_in}$$

$$y2 = \overline{\overline{b\_in}} \cdot a\_in$$

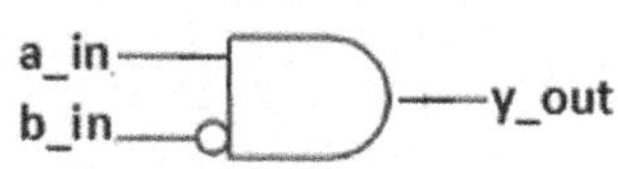

Fig. 2.15 Custom-gate

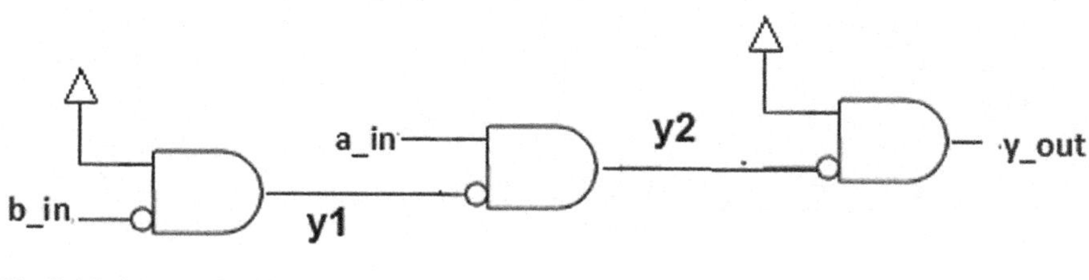

Fig. 2.16 2-input NAND using custom gate

$$y2 = a\_in \cdot b\_in$$

$$y\_out = 1 \cdot \overline{y2}$$

$$y\_out = \overline{a\_in \cdot b\_in}$$

Now as by using the above custom gate, we can implement the NAND which is universal gate, and we can treat this custom logic gate as universal gate.

2.4.2 *Exercise 2: Design Using the MUX*

Using the minimum number of 2:1 multiplexers, design the XOR gate.

Solution: Now the better strategy is to use the truth table of XOR gate (Table 2.14). As shown, the number of entries is 4, and as a_in = 0 for first two entries and a_in = 1 for next two entries to realize XOR using the minimum number of 2:1 mux, let us compare b_in with y_out of XOR gate.

As shown (Table 2.15), for **a_in = 0, y_out = b_in and for a_in = 1, y_out =** $\overline{\textbf{b\_in}}$.

The XOR logic using two 2:1 multiplexers is shown in Fig. 2.17.

Table 2.14 XOR gate truth-table

a_in	b_in	y_out
0	0	0
0	1	1
1	0	1
1	1	0

Table 2.15 Truth-table for realization of the XOR gate

a_in	y_out
0	b_in
1	$\overline{\text{b\_in}}$

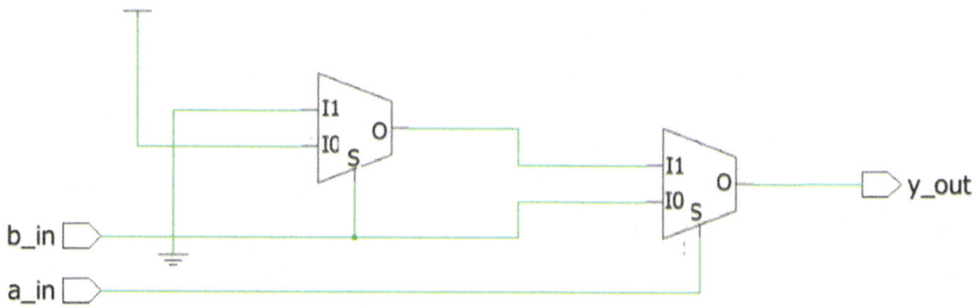

Fig. 2.17 XOR gate using 2:1 MUX

2.4.3 Exercise 3: Design Using MUX

Using the minimum number of 2:1 multiplexers, design the XNOR gate.

Solution: Now the better strategy is to use the truth table of XNOR gate (Table 2.16). As shown, the number of entries is 4 and as a_in = 0 for first two entries, and a_in = 1 for next two entries to realize XNOR using the minimum number of 2:1 mux, let us compare b_in with y_out of XNOR gate.

Table 2.16 2-input XNOR gate

a_in	b_in	y_out
0	0	1
0	1	0
1	0	0
1	1	1

Table 2.17 XNOR entries to realize logic using multiplexers

a_in	y_out
0	$\overline{\text{b\_in}}$
1	b_in

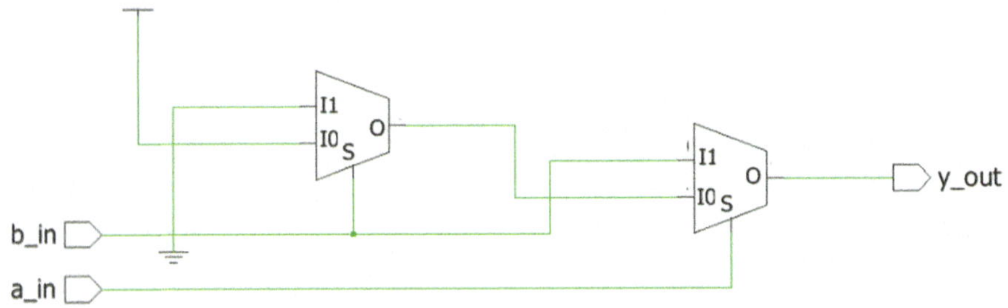

Fig. 2.18 XNOR gate using 2:1 multiplexers

As shown (Table 2.17), for **a_in = 0, y_out = $\overline{\text{b\_in}}$ and for a_in = 1, y_out = b_in**.

The XNOR logic using two 2:1 multiplexers is shown in Fig. 2.18.

2.4.4 *Exercise 4: Design Using Custom Gates*

Using the following custom gate, design the 2-input NOR gate (Fig. 2.19).

Solution: Let us use the understanding on the universal logic gates, and by using the minimum number of these gates, let us design the 2-input NOR gate.

As NOT of OR is NOR. Let us implement the OR and then NOT of OR (Fig. 2.20).

$$y1 = 0 + \overline{\text{a\_in}}$$

$$y1 = \overline{\text{a\_in}}$$

$$y2 = \overline{\overline{\text{a\_in}}} + \text{b\_in}$$

Fig. 2.19 Custom gate

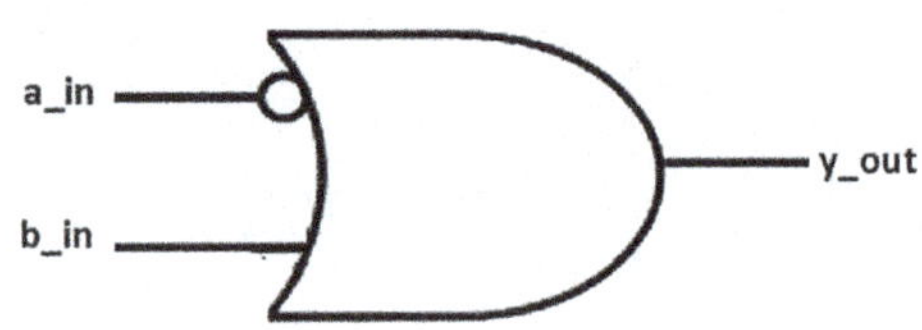

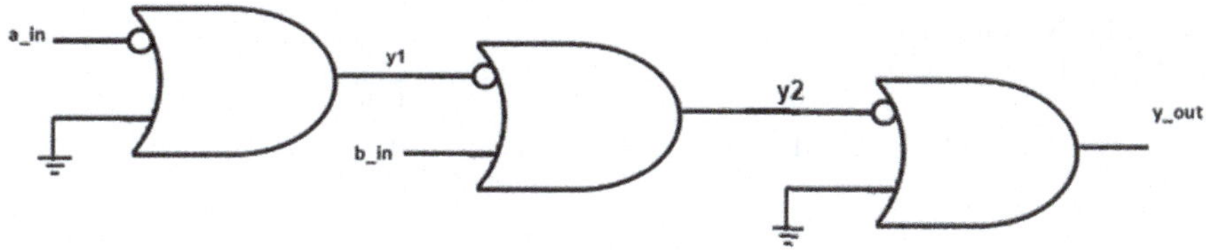

Fig. 2.20 2-input OR using the application-specific gates

$$y2 = a\_in + b\_in$$

$$y\_out = \overline{a\_in + b\_in}$$

Now as by using the above custom gate, we can implement the NOR which is universal gate, we can treat this custom logic as universal gate.

2.4.5 Exercise 5: Optimization Exercise

Using the minimum number of 2:1 multiplexers, design the NOR gate.

Solution: Now the better strategy is to use the truth table of NOR gate (Table 2.18). As shown, the number of entries is 4, and as a_in = 0 for first two entries and a_in = 1 for next two entries to realize NOR using the minimum number of 2:1 mux, let us compare b_in with y_out of NOR gate.

As shown (Table 2.19), for **a_in = 0, y_out =** $\overline{\textbf{b\_in}}$ **and for a_in = 1, y_out = 0**.

Table 2.18 Truth-table of NOR gate

a_in	b_in	y_out
0	0	1
0	1	0
1	0	0
1	1	0

Table 2.19 NOR to realize using 2:1 mux

a_in	y_out
0	$\overline{\text{b\_in}}$
1	0

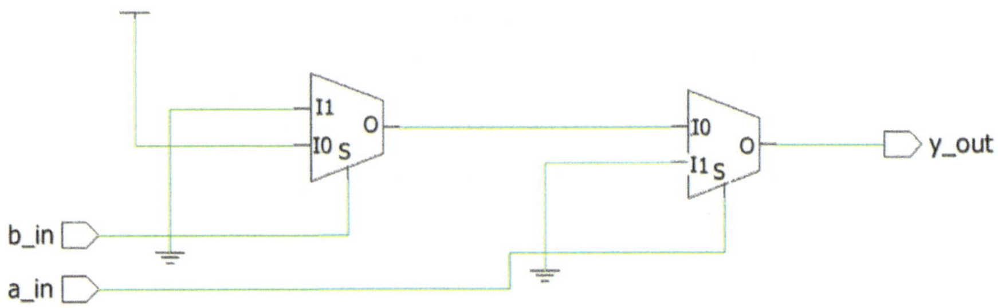

Fig. 2.21 2-input NOR using only 2:1 multiplexers

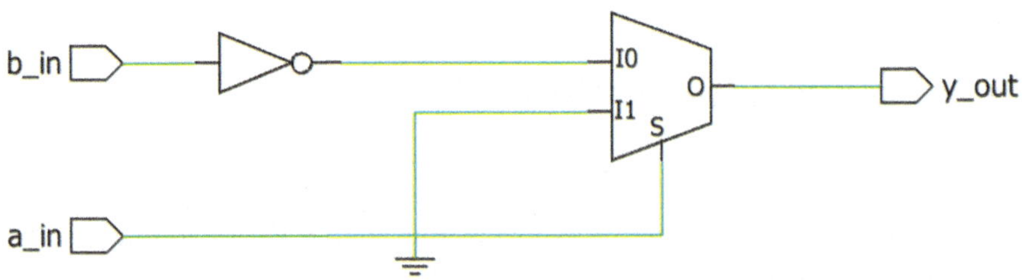

Fig. 2.22 NOR using NOT and 2:1 MUX

The NOR logic using two 2:1 multiplexers is shown in Fig. 2.21.

The input multiplexer to get complement of the b_in can be optimized using the NOT gate. Hence to implement the 2-input NOR (Fig. 2.22), we need to have single 2:1 mux and NOT gate.

2.4.6 Exercise 7: Design Using the MUX

Using the minimum number of 2:1 multiplexers, design the clock muxing. Consider for the clk_select = 1 output should be clk_1 and for clk_select = 0 output should be clk_2.

Solution: Now as only two clock inputs, let us tabulate the specification (Table 2.20). As shown, the number of entries is 2 and as clk_select = 0 the y_out = clk_2. For clk_select = 1, the y_out = clk_1.

Now use the select line s = clk_select and inputs as I1 = clk_1, I0 = clk_2 to get clk_out from single 2:1 mux. The design is shown in Fig. 2.23.

Table 2.20 Clock muxing

clk_select	y_out
0	clk_2
1	clk_1

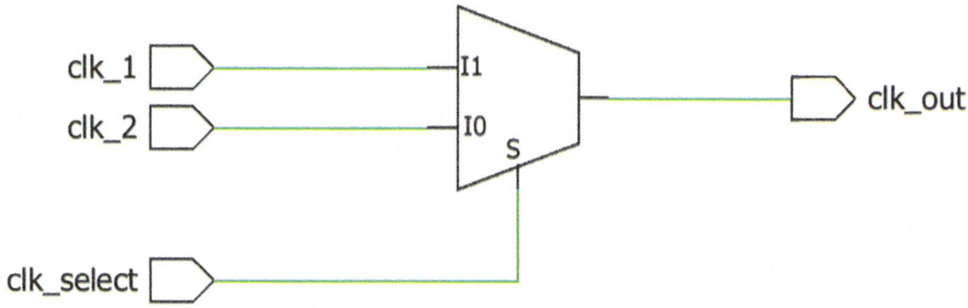

Fig. 2.23 Clock muxing

2.4.7 Exercise 8: Design Using MUX

Using the minimum number of suitable multiplexer, design the following SOP function.

$$f(\text{sel\_in}[1], \text{sel\_in}[0]) = \sum m(1, 2)$$

Solution: As given, the Boolean function is Sum of Product (SOP) (Table 2.21). As shown, the number of entries is 4, and output f is equal to logic 1 for min terms that is 1, 2. For other combinations, that is for decimal input 0 and 3 output is logic 0.

This can be implemented using pull up (Vdd) or pull down (Vss) at inputs of single 4:1 mux (Fig. 2.24). The select lines of the multiplexer are sel_in[1], sel_in[0], and output is y_out. The inputs I0 = 0, I1 = 1, I2 = 1 and I3 = 0.

Use the strategy in the Exercise 2 to implement using minimum number of 2:1 multiplexers.

Table 2.21 Truth-table for the given function

sel_in[1]	sel_in[0]	f = y_out
0	0	0
0	1	1
1	0	1
1	1	0

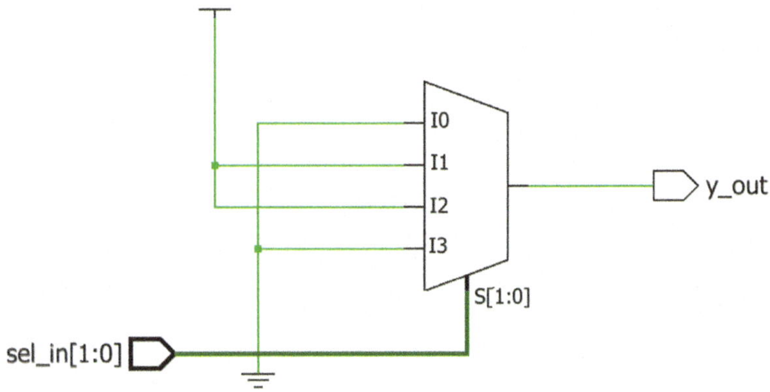

Fig. 2.24 XOR using 4:1 MUX

2.4.8 Exercise 9: Design Using Custom Gates

Using the minimum number of suitable multiplexer, design the following SOP function.

$$f(\text{sel\_in}[1], \text{sel\_in}[0]) = \sum m(0, 3)$$

Solution: As given, the Boolean function is Sum of Product (SOP) (Table 2.22). As shown, the number of entries is 4, and output f is equal to logic 1 for min terms that is 0, 3. For other combinations, that is 1 and 2 output is logic 0.

This can be implemented using pull up (Vdd) or pull down (Vss) at inputs of single 4:1 mux (Fig. 2.25). The select lines of the multiplexer are sel_in[1], sel_in[0], and output is y_out. The inputs should be I0 = 1, I1 = 0, I2 = 0 and I3 = 1.

Use the strategy in the Exercise 2 to implement the design using minimum number of 2:1 multiplexers.

Table 2.22 Truth-table for the given function

sel_in[1]=a_in	sel_in[0]=b_in	f = y_out
0	0	1
0	1	0
1	0	0
1	1	1

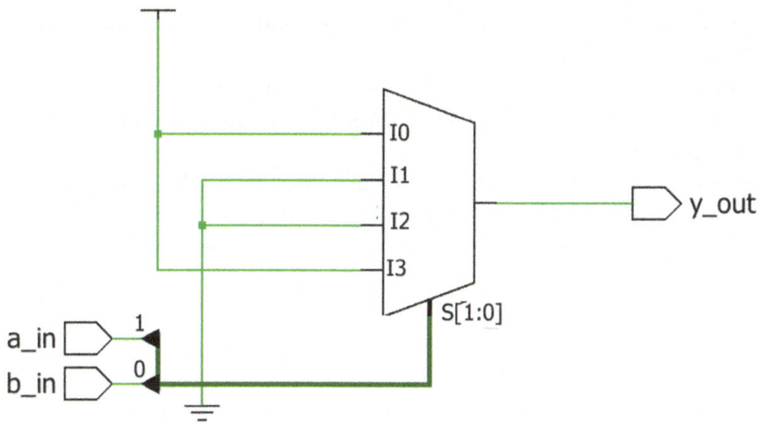

Fig. 2.25 XNOR using 2:1 MUX

2.5 Applications and Use in VLSI Context

Following are the applications of the mux-based logic and universal logic gates in the VLSI context.

In the VLSI design context, the following are goals of the designers:

1. ***Understand the Boolean function to be realized and have strategy in place to use the minimum logic gates.***
2. ***Use the minimum number of universal logic gates to implement the combinational logic.***
3. ***Use the minimum number of multiplexers and try to avoid the cascade stages.***
4. ***While implementing the design using multiplexers, avoid the use of the priority logic. Use the mux based logic for pin multiplexing and combinational logic realization.***

2.6 Important Takeaways

Following are few of the important points to conclude this chapter:

1. NAND and NOR are universal logic gates.
2. Using 2-input four NAND gates, 2-input XOR gate is realized and uses only 4 NAND gates.

3. The minimum number of 2-input NOR gates required to realize the 2-input XNOR gate is 4.
4. Multiplexers are used to realize the Boolean functions.
5. The minimum number of 2:1 mux required to realize the XOR, XNOR, NOR and NAND gates is equal to 2.
6. Only single 2:1 MUX is required to realize the NOT, OR, AND gate.
7. Multiplexers are used in the pin muxing and clock muxing.

Chapter 3
Combinational Design Resources

Various combinational resources and the design techniques are useful to design the arithmetic and other processing logic.

The chapter discusses the various code converters, combinational design resources and the arithmetic resources. The design techniques discussed in this chapter are useful to design the combinational or glue logic. The chapter even focuses on the various performance improvement techniques and their use to design the combinational logic.

3.1 Code Converters

Most of us are familiar with the various code converters those are useful in the design. Most of the time, we use the binary-to-gray and gray-to-binary code converters. As in the two successive gray codes, only one bit changes; these codes are used to improve the overall power for the design. The main application is use of these code converters in the multiple clock domain and in the FSM designs. Due to lower toggling rate, they are useful to improve the overall power for the design.

3.1.1 Three-Bit Binary-to-Gray Code Converter

The 3-bit binary-to-gray code truth-table (Table 3.1) gives relationship between the binary number and gray number.

The design of the 3-bit binary-to-gray code converter is discussed here. As eight entries of the binary and gray codes, let us use the three-variable K-map to deduce the equations for g2_out, g1_out, g0_out.

As a combinational logic, g2_out, g1_out and g0_out is function of the b2_in, b1_in and b0_in.

V. Taraate, *Digital Design Techniques and Exercises*,
https://doi.org/10.1007/978-981-16-5955-3_3

Table 3.1 3-bit binary and gray code

3-bit binary code	3-bit gray code
000	000
001	001
010	011
011	010
100	110
101	111
110	101
111	100

1. Due to group of four 1's, the logic expression using the K-map for g2_out (Fig. 3.1) is

$$g2\_out = b2\_in$$

2. Let us group the terms as shown in the K-map (Fig. 3.2). The logic expression is

$$g1\_out = \overline{b2\_in}.\, b1\_in + \overline{b1\_in}.b2\_in$$
$$g1\_out = b2\_in \oplus b1\_in$$

3. Let us group the terms as shown in the K-map (Fig. 3.3). The logic expression is

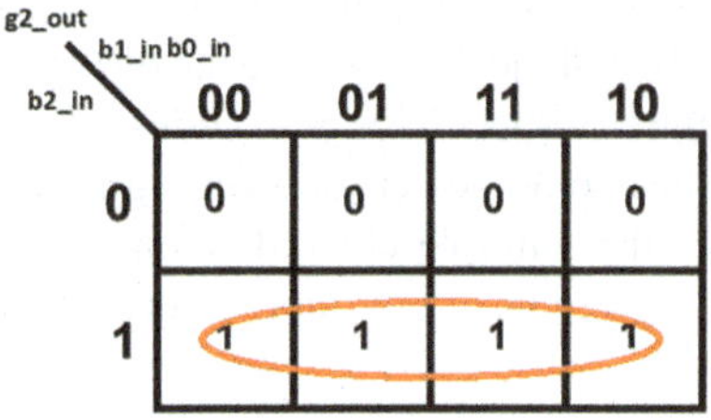

Fig. 3.1 K-map for g2_out

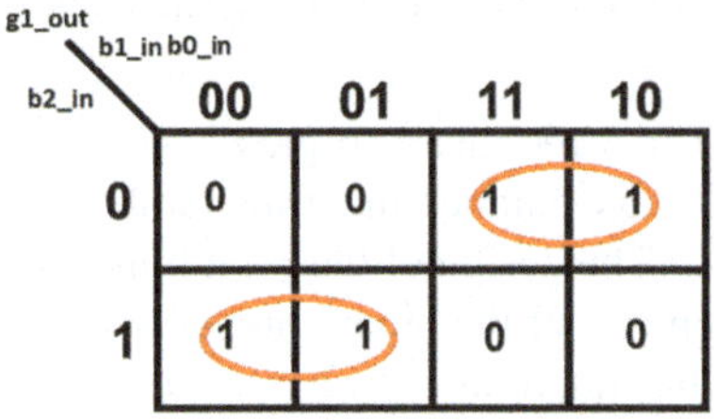

Fig. 3.2 K-map for g1_out

Fig. 3.3 K-map for g0_out

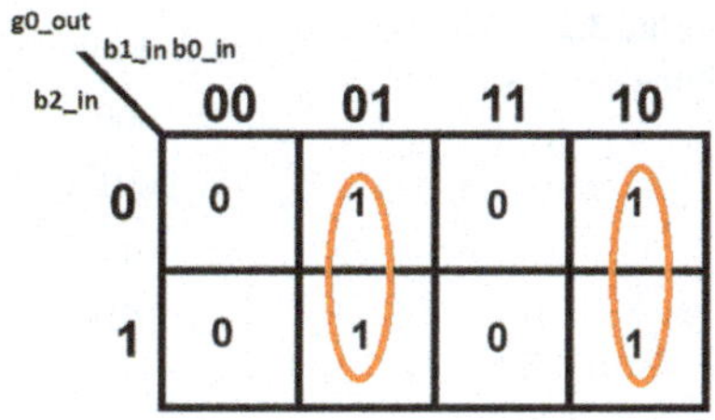

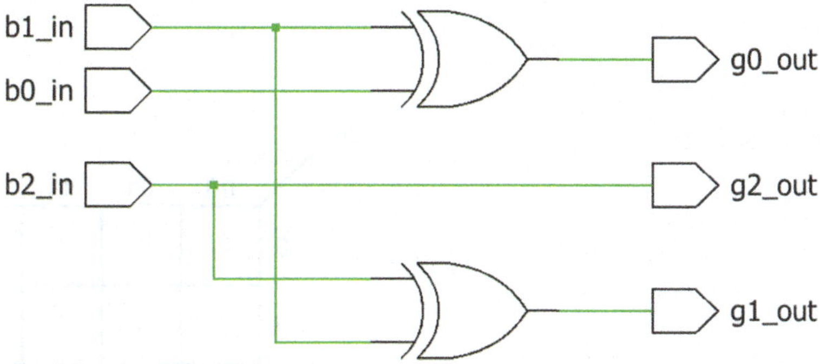

Fig. 3.4 3-bit binary-to-gray code converter

$$g0\_out = \overline{b1\_in}.b0\_in + \overline{b0\_in}.b1\_in$$
$$g0\_out = b1\_in \oplus b0\_in$$

So to implement the 3-bit binary-to-gray code converter we need to have the 2-XOR gates. The design of the 3-bit binary-to-gray code converter is shown in Fig. 3.4.

3.1.2 3-Bit Gray-to-Binary Code Converter

The 3-bit gray-to-binary code converter truth-table (Table 3.2) describes the relationship between the 3-bit gray code (**g2_in, g1_in and g0_in**) and binary code (**b2_out, b1_out, b0_out**).

The design of the 3-bit gray-to-binary code converter is discussed here. As eight entries of the gray and binary code, let us use the three-variable K-map to deduce the equations for b2_out, b1_out, **b0_out**.

As a combinational logic g2_out, g1_out and g0_out is function of the b2_in, b1_in and b0_in.

1. Due to group of four 1's, the logic expression using the K-map for b2_out (Fig. 3.5) is

Table 3.2 3-bit gray and binary codes

3-bit gray code	3-bit binary code
000	000
001	001
011	010
010	011
110	100
111	101
101	110
100	111

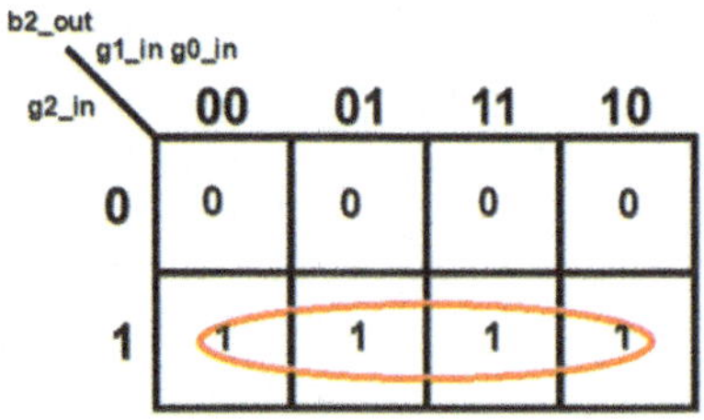

Fig. 3.5 K-map for the b2_out

$$b2\_out = g2\_in$$

2. Let us group the terms as shown in the K-map (Fig. 3.6). The logic expression is

$$b1\_out = \overline{g2\_in}.g1\_in + \overline{g1\_in}.g2\_in$$
$$b1\_out = g2\_in \oplus g1\_in$$

3. Let us group the terms as shown in the K-map (Fig. 3.7). The logic expression is

$$b0\_out = \overline{g2\_in}.\overline{g1\_in}.g0\_in + \overline{g2\_in}.\overline{g0\_in}.g1\_in$$
$$+ g2\_in.\overline{g1\_in}.\overline{g0\_in} + g2\_in.g1\_in.g0\_in$$

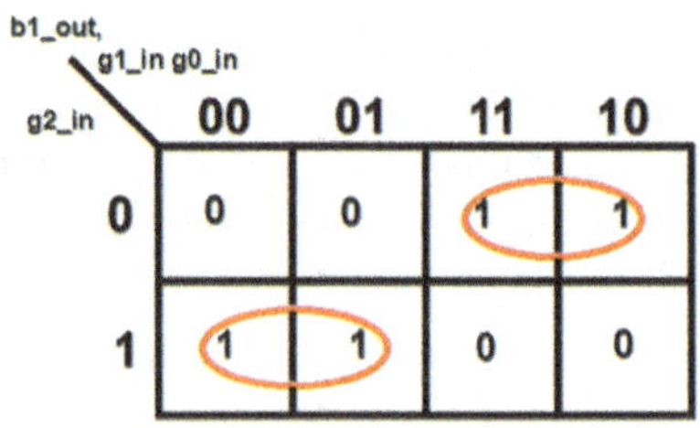

Fig. 3.6 K-map for the b1_out

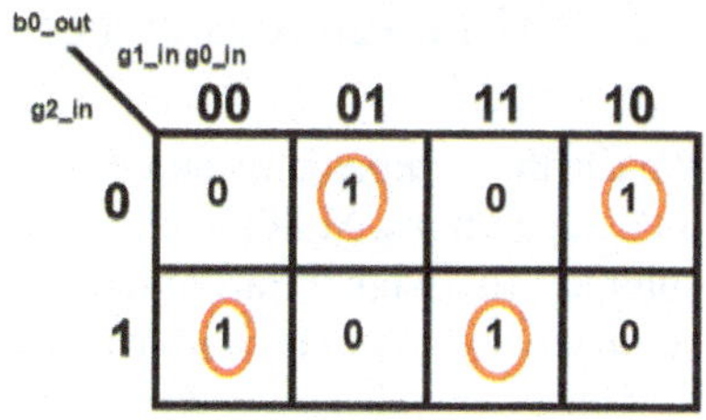

Fig. 3.7 K-map for the b0_out

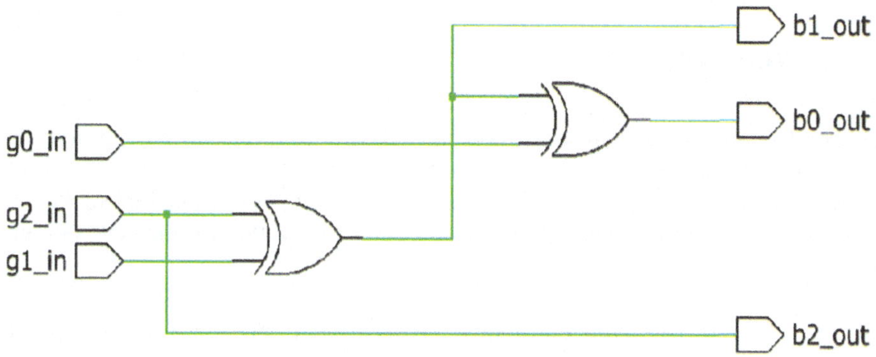

Fig. 3.8 3-bit gray-to-binary code converter

$$\begin{aligned} \text{b0\_out} &= \overline{\text{g2\_in}}.\left(\overline{\text{g1\_in}}.\text{g0\_in} + \overline{\text{g0\_in}}.\text{g1\_in}\right) \\ &\quad + \text{g2\_in}.\left(\overline{\text{g1\_in}}.\overline{\text{g0\_in}} + \text{g1\_in}.\text{g0\_in}\right) \\ \text{b0\_out} &= \text{g2\_in} \oplus \text{g1\_in} \oplus \text{g0\_in} \end{aligned}$$

So, to implement the 3-bit gray-to-binary code converter, we need to have the 2-XOR gates. The design of the 3-bit binary-to-gray code converter is shown in Fig. 3.8.

In the VLSI design context, the following are use of the code converters

1. The binary-to-gray code converters are used in multiple clock domain designs.
2. The gray pointers and gray counters are used in the multiple clock domain designs as in the two successive gray codes, only one bit changes.
3. If gray values are passed from one of the clock domain to another clock domain, then to get the equivalent binary value, the gray-to-binary code converters are useful.

3.2 Arithmetic Resources

We use the arithmetic resources such as adders and subtractors to perform the arithmetic operations. Multiplication can be successive addition, and division is shift and subtract. To design the arithmetic and logic unit, we can think of using the adders and subtractors with other combinational elements such as logic gates and multiplexers.

This section discusses the arithmetic resources and their role in the digital design.

3.2.1 Half-Adder

As most of us are familiar with the basic resource to perform the addition, the resource is half-adder. It performs the addition of the a_in, b_in to generate the result as sum_out, carry_out. The truth table of the half-adder is shown in Table 3.3.

Now let us use the truth table to get the Boolean function for the sum_out and carry_out.

$$\text{sum\_out(a\_in, b\_in)} = \sum m(1, 2)$$

The K-map of the sum_out denotes the entries for the Boolean function and shown in Fig. 3.9.

The Boolean equation is derived from the number of 1's which are circled in the K-map.

Table 3.3 Half-adder truth-table

a_in	b_in	sum_out	carry_out
0	0	0	0
0	1	1	0
1	0	1	0
1	1	0	1

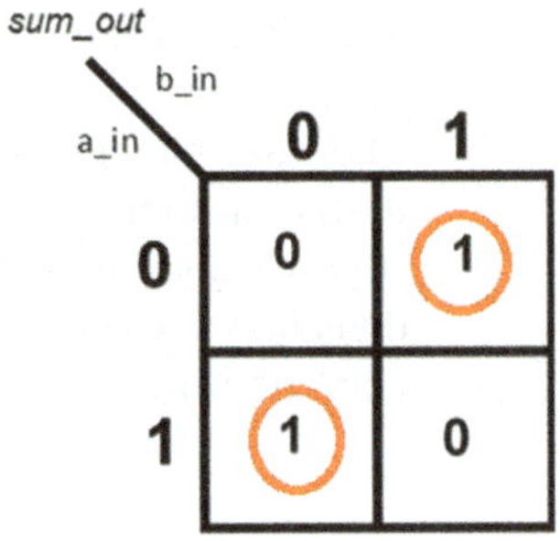

Fig. 3.9 K-map for the sum_out

Fig. 3.10 K-map for the carry_out

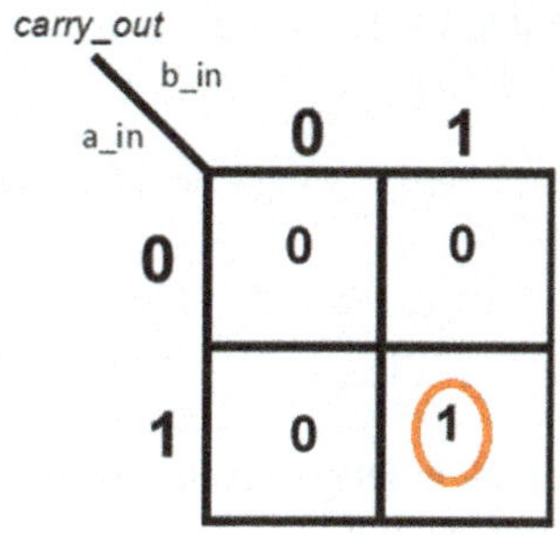

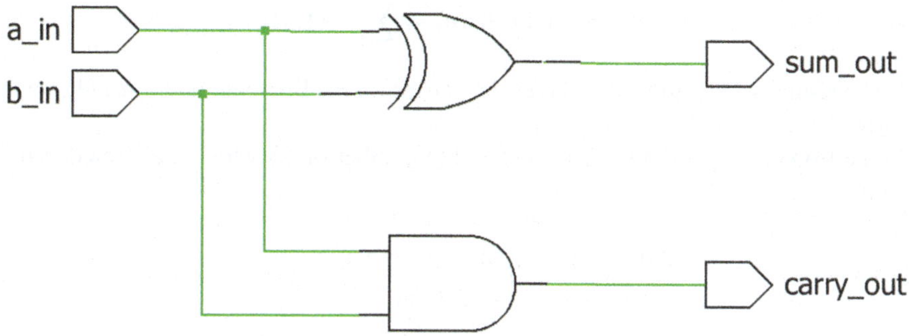

Fig. 3.11 Half-adder using logic gates

$$\text{sum\_out} = \overline{\text{a\_in}} \,.\, \text{b\_in} + \overline{\text{b\_in}} \,.\, \text{a\_in}$$
$$\text{sum\_out} = \text{a\_in} \oplus \text{b\_in}$$

The K-map of the carry_out denotes the entries for the Boolean function and shown in Fig. 3.10.

The Boolean equation is derived from the number of 1's which are circled in the K-map.

$$\text{carry\_out (a\_in, b\_in)} = \sum m\,(3)$$
$$\text{carry\_out} = \text{a\_in} \,.\, \text{b\_in}$$

The design of the half-adder is shown in Fig. 3.11.

3.2.2 Half-Subtractor

As name indicates, the half-subtractor is used to perform the subtraction of a_in, b_in, and it generates the output as diff_out, borrow_out. The truth table of the half-subtractor is shown in Table 3.4.

Table 3.4 Half-subtractor truth-table

a_in	b_in	diff_out	borrow_out
0	0	0	0
0	1	1	1
1	0	1	0
1	1	0	0

Now let us use the truth table to get the Boolean function for the diff_out and borrow_out.

$$\text{diff\_out}(\text{a\_in},\ \text{b\_in}) = \sum m\,(1,\ 2)$$

The K-map of the diff_out denotes the entries for the Boolean function and shown in Fig. 3.12.

The Boolean equation is derived from the number of 1's which are circled in the K-map.

$$\text{diff\_out} = \overline{\text{a\_in}}\,.\,\text{b\_in} + \overline{\text{b\_in}}\,.\,\text{a\_in}$$
$$\text{diff\_out} = \text{a\_in} \oplus \text{b\_in}$$

The K-map of the borrow_out denotes the entries for the Boolean function and shown in Fig. 3.13.

The Boolean equation is derived from the number of 1's which are circled in the K-map.

$$\text{borrow\_out}\ (\text{a\_in},\ \text{b\_in}) = \sum m\,(1)$$
$$\text{borrow\_out} = \overline{\text{a\_in}}\,.\,\text{b\_in}$$

The design of the half-subtractor is shown in Fig. 3.14.

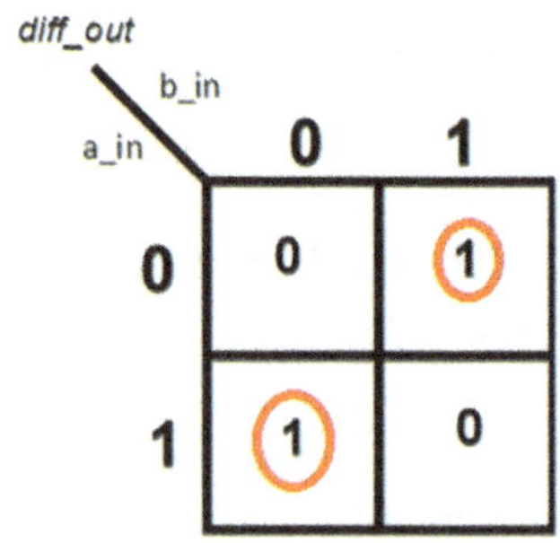

Fig. 3.12 K-map for the diff_out

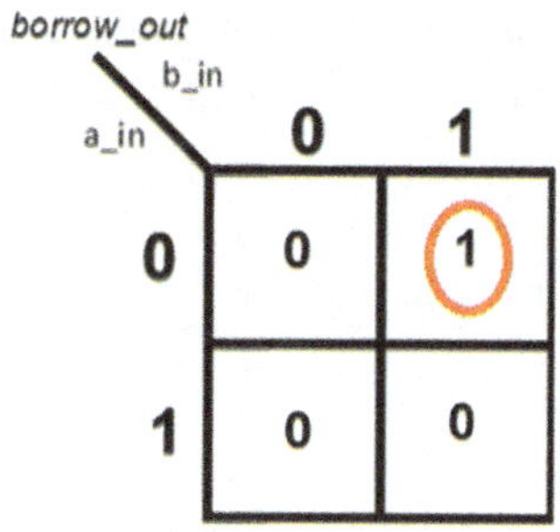

Fig. 3.13 K-map for the borrow_out

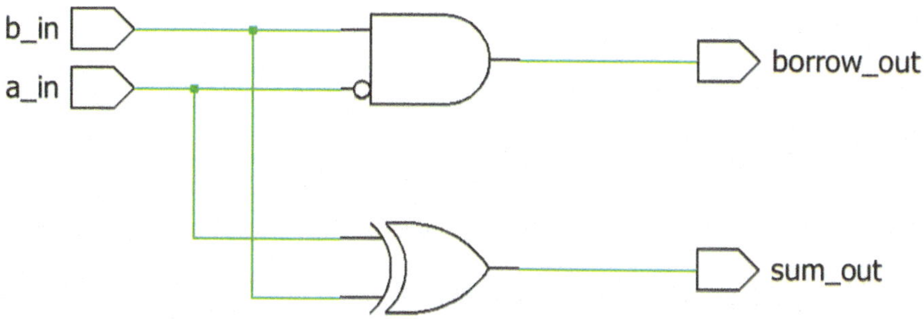

Fig. 3.14 Half-subtractor using logic gates

In the VLSI design context, the following are few of the important points while using adders.

1. The adders are used to perform the additions, and designers should avoid the use of the cascade adders.
2. The subtraction is 2' complement addition, and to perform the addition and subtraction, try to use the common resources.
3. Adders consume more area as compared to multiplexers; so use less number of adders and more number of multiplexers.

3.2.3 Full-Adder

The full-adder uses the inputs a_in, b_in and carry input c_in and performs the addition and generates the result as sum_out, carry_out (Table 3.5).

The resources needed to implement the full-adder.

1. Half-adder
2. OR gate.

Table 3.5 Truth-table of full-adder

a_in	b_in	c_in	sum_out	carry_out
0	0	0	0	0
0	0	1	1	0
0	1	0	1	0
0	1	1	0	1
1	0	0	1	0
1	0	1	0	1
1	1	0	0	1
1	1	1	1	1

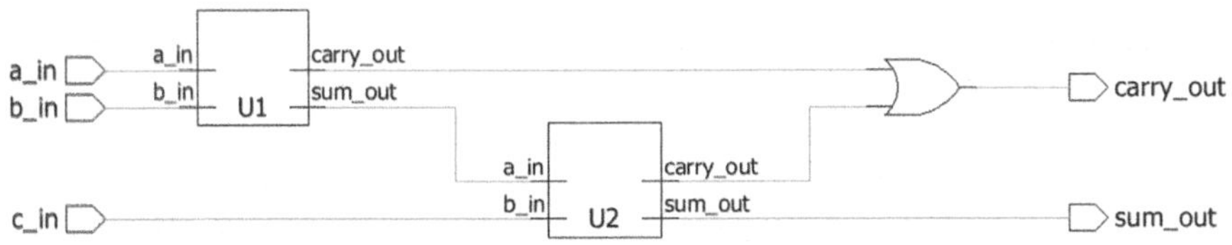

Fig. 3.15 Full-adder using half-adders and OR gate

The full-adder can be designed using the half-adders connected in cascade. To generate the carry output, use the OR gate. The design of the full-adder using the two half-adders and OR gate is shown in Fig. 3.15.

3.3 Use of Arithmetic Resources in the Design

The design scenarios while using the arithmetic resources are discussed in this section. Consider the design requirement to perform the addition of a_in, b_in, c_in, d_in.

$$y\_\text{out} = \text{a\_in} + \text{b\_in} + c\_\text{in} + d\_\text{in}$$

We can think of the group of two inputs, and then, we can perform the addition

$$y\_\text{out} = (\text{a\_in} + \text{b\_in}) + (c\_\text{in} + d\_\text{in})$$

This strategy will use the three adders. The design is shown in Fig. 3.16. As shown if each adder has 1 ns delay, then the overall delay to get the result is 2 ns.

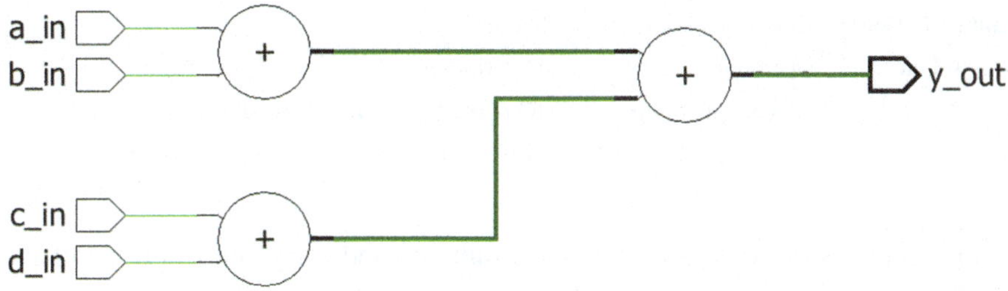

Fig. 3.16 Design using adders

3.4 Design Using Arithmetic Resources and Control Elements

The design scenarios while using the arithmetic resources and multiplexers are discussed in this. Consider the design scenario that for the control input logic 1, the operation to be performed is ADD (a_in, b_in) and for control input logic 0 the design should perform the operation ADD (c_in, d_in).

Now what should be our strategy to design the arithmetic operation circuit?

We can use the adder as resource and multiplexer to select the operation a_in + b_in, c_in + d_in depending on the status of the control input. The operations are documented in Table 3.6.

The design for the arithmetic operation is shown in Fig. 3.17. As shown at the input side, the two-adders are used and they perform the operation on (a_in, b_in),

Table 3.6 Design for arithmetic operation

Control_in	Operation	Description
0	ADD (c_in, d_in)	Perform the addition of c_in, d_in
1	ADD (a_in, b_in)	Perform the addition of a_in, b_in

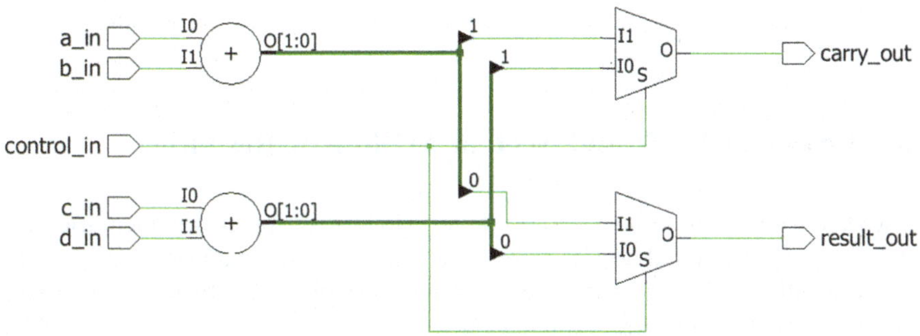

Fig. 3.17 Design using adders and multiplexers

Table 3.7 Entries with the goal of resource sharing

Control_in	Operation	Description
0	ADD (c_in, d_in)	IO of each 2:1 mux is c_in, d_in respectively
1	ADD (a_in, b_in)	I1 of each 2:1 mux is a_in, b_in respectively

(c_in, d_in), respectively. To select the result_out and carry_out from one of the operations, the multiplexers are used at the output side.

The design has more area as it needs the two-adders and two-multiplexers. Refer the optimization goals and strategy section to optimize the arithmetic resources using the resource sharing concept.

The arithmetic resources for the processor-based designs and the design optimization strategies are discussed in the following section.

3.5 Optimization Goals

For the design of the arithmetic or logic circuits, the main optimization goals are:

1. Optimization for the area
2. Optimization for the speed.

Consider the design which is discussed in Sect. 3.4, and we need to optimize the design. We can use the common resource as adder at output and the tree of mux at the input side. The modified operations to have resource sharing are shown in Table 3.7.

As shown if we use the IO input of each input multiplexer as c_in, d_in respectively, then for the control_in = 0 the adder performs the operation c_in + d_in.

For the I1 input of each input multiplexer as a_in, b_in respectively, then for the control_in = 1, the adder performs the operation a_in + b_in.

The design is shown in Fig. 3.18. As shown the common resource single adder is used, and this technique improves the speed and the area for the design. Refer Chap. 4 for the ALU design and optimization of the data and control paths.

3.6 Processor Logic and Need of Arithmetic Resources

As we know that in the system design, we need to have the processor, memories and IO devices. As the digital designer, we should have the strong understanding of the processor functional blocks. If we consider the 16-bit processor, then we can use the arithmetic resources to perform the operations.

1. 16-bit addition

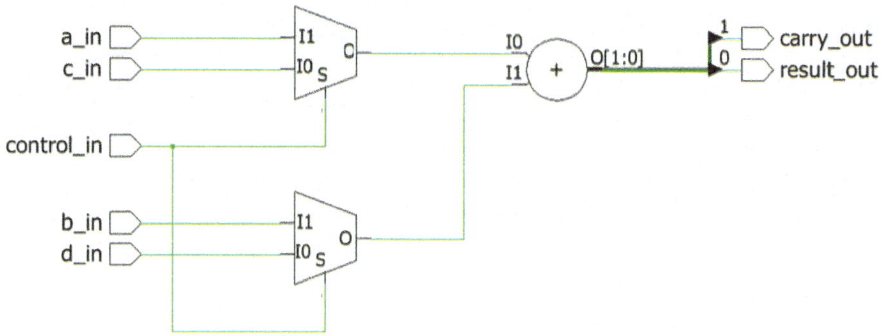

Fig. 3.18 Use of the common resources at output

2. 16-bit subtraction
3. 16-bit multiplication
4. Combinational shifting.

To design these operations, our goal is to use the minimum number of the arithmetic resources to have the minimum area and to have less propagation delay. For more details about the architecture design and other advanced design concepts, refer Chaps. 10–12.

3.7 Exercises

The exercises on the use of the arithmetic resources and optimization are discussed in this section. Even the exercises are useful to understand the speed and area for the design.

3.7.1 Exercise 1: Cascade Versus Parallel Logic

For the following design (Fig. 3.19), find the propagation delay. Consider propagation delay of each gate is 1 ns. Design the parallel logic to improve the delay of the design.

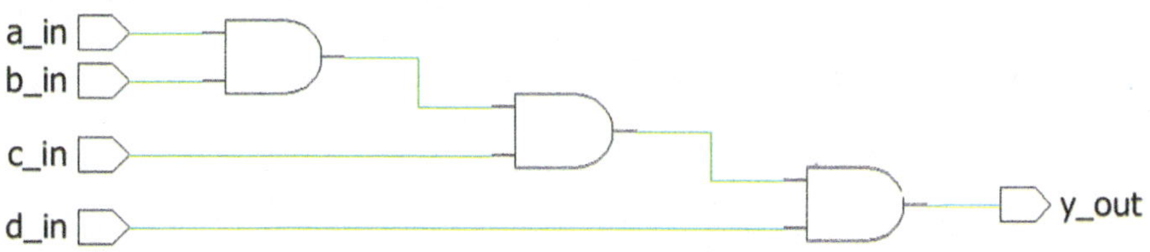

Fig. 3.19 Cascade logic

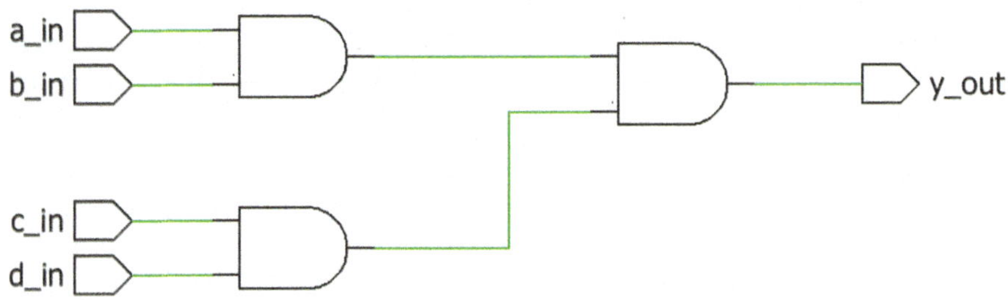

Fig. 3.20 Parallel logic

Solution: Each AND gate has delay of 1 ns. So, for the Boolean equation **y_out = (a_in. b_in. c_in. d_in)** the propagation delay is 3 ns. We can use the parallel logic by grouping the terms. The strategy is shown below

$$y\_out = (\text{a\_in} \ . \ \text{b\_in}) \ . \ (c\_\text{in} \ . \ d\_\text{in})$$

The design shown in Fig. 3.20 uses three-AND gates, but as at input side the AND gates are used in parallel; the propagation delay is 2 * tpff = 2 * 1 ns = 2 ns.

3.7.2 Exercise 2: Delay of the Design

For the following design (Fig. 3.21), find the propagation delay. Consider propagation delay of each adder is 1 ns.

Solution: Each adder has delay of 1 ns. So, for the Boolean equation **y_out = (a_in + b_in + c_in + d_in)** the resources used are three adders. Each adder has propagation delay of 1 ns, so the overall propagation delay is 3 * 1 ns = 3 ns.

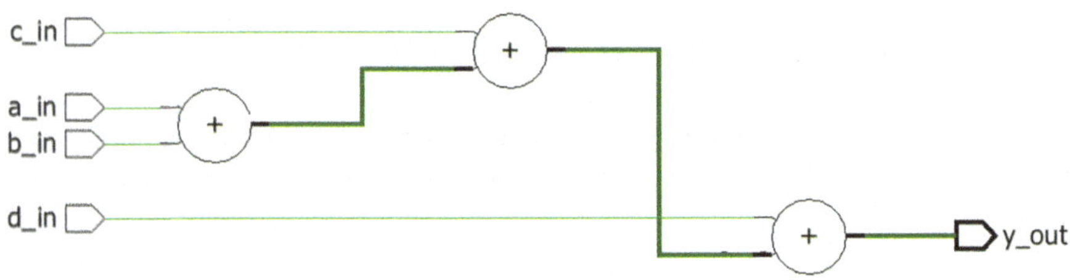

Fig. 3.21 Adder in the cascade stages

Fig. 3.22 Odd number of NOT gates in cascade

3.7.3 *Exercise 3: Speed*

For the following design (Fig. 3.22), find the propagation delay. Consider propagation delay of each gate is 1 ns.

Solution: Each gate has delay of 1 ns. So, for the logic shown in the figure, NOT gate delay is 1 ns and buffer (Even NOT in cascade) delay is also 1 ns. So the overall propagation delay is 2 * 1 ns = 2 ns.

3.7.4 *Exercise 4: Design to perform the Addition and Subtraction*

Design the logic using the minimum resources to perform the following operation shown in (Table 3.8).

Solution: As described in the table the control_in = 0 the operation is addition of a_in, b_in and for control_in = 1 the operation is subtraction of a_in, b_in. We can use resources as

1. Adder
2. Subtractor
3. Multiplexers for result_out and carry or borrow out.

The design is shown in Fig. 3.23.

Table 3.8 Adder–subtractor

Control_in	Operation	Description
0	ADD (a_in, b_in)	Perform the subtraction of a_in, b_in
1	SUB (a_in, b_in)	Perform the addition of a_in, b_in

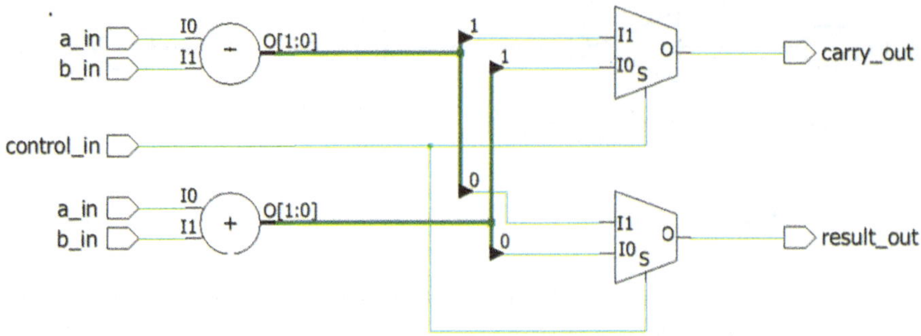

Fig. 3.23 Addition and subtraction without resource optimization

3.7.5 *Exercise 4: Design with the Goal to Use Resource Sharing*

Design the logic using the resource optimization to perform the following operation shown in (Table 3.9).

Solution: As the goal is to use the common resources, we can think of the use of the resource sharing. That is perform the subtraction using 2's complement addition. The operations are documented in Table 3.10.

We can use the single adder, and we can control the input of adder depending on the control input. The entries are shown in Table 3.11.

Table 3.9 Addition and subtraction table

Control_in	Operation	Description
1	SUB (a_in, b_in)	Perform the subtraction of a_in, b_in
0	ADD (a_in, b_in)	Perform the addition of a_in, b_in

Table 3.10 Resource sharing strategy

Control_in	Operation	Description
1	SUB (a_in, b_in)	a_in-b_in = a_in + $\overline{\textbf{\textit{b\_in}}}$+1
0	ADD (a_in, b_in)	a_in + b_in = a_in + b_in + 0

Table 3.11 Adder inputs depending on the control input status

Control_in	*x*	*y*
1	a_in	$\overline{\text{b\_in}}$
0	a_in	b_in

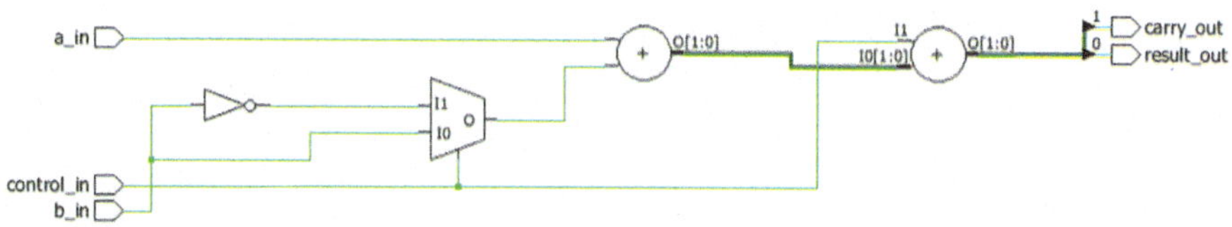

Fig. 3.24 Addition and subtraction using minimum resources

The design which uses the adder as common resource and the combinational logic using 2:1 mux is shown in Fig. 3.24. The design has better area and better speed, power as only one operation is performed at a time.

3.8 Important Takeaways

Following are few of the important points to conclude this chapter.

1. In the two successive gray codes, only one bit changes, and they are used to improve the overall power for the design.
2. Binary-to-gray and gray-to-binary codes are designed using the XOR gates.
3. The adders are used to perform the addition and subtraction. For the subtraction, we can use the 2's complement addition.
4. Cascade logic increases the propagation delay.
5. Parallel logic is useful to reduce the overall propagation delay for the design.
6. The resource sharing is useful to improve the design performance.

Chapter 4
Case Study: ALU Design

The design for ALU should use the less arithmetic resources and should have better data and control path design.

In this chapter, let us try to use the combinational resources and arithmetic elements to design the digital circuit. The objective is to optimize the design to have the least area and maximum speed. The chapter discusses about the basics of the instruction processing and the optimization for the area and speed.

4.1 Design Specifications and Their Role

Consider the design of the arithmetic operations that is, addition and subtraction. The processing unit performs these two operations depending on the control input status. When control_in is logic 0, it performs the addition operation, and when the control_in input is logic 1, it performs the subtraction operation.

What should be out thought process?

We should think with reference to the given functional specifications. The few important points are documented below

1. How many number of inputs and outputs?
2. What is the size of the data inputs and outputs?
3. What are the different operations and what elements we can use?
4. Can we share the common resources?

By considering all the above points, we can design the logic. For better area, these two operations are documented in Table 4.1.

Now, the resource required for the addition is full-adder and for the subtraction is full-subtractor. To optimize for the logic, we can perform the subtraction using 2's complement addition.

V. Taraate, *Digital Design Techniques and Exercises*,
https://doi.org/10.1007/978-981-16-5955-3_4

Table 4.1 Addition–subtraction operations

Control input	Operation	Description
0	ADD(a_in, b_in)	Addition of the a_in, b_in
1	SUB(a_in, b_in)	Subtraction of a_in, b_in

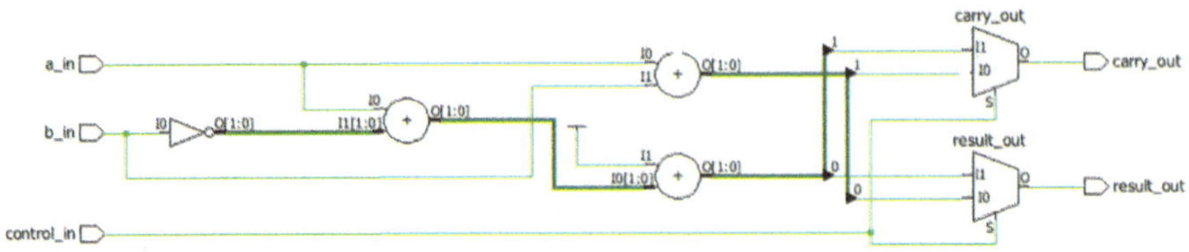

Fig. 4.1 Addition and subtraction

$$\text{ADD(a\_in, b\_in)} = \text{a\_in} + \text{b\_in} + 0;$$
$$\text{SUB (a\_in, b\_in)} = \text{a\_in} - \text{b\_in} = \text{a\_in} + \sim \text{b\_in} + 1;$$

Where ($\sim$) is pronounced as NOT.

Now to select from one of the operations, we can think of using the multiplexer logic. The selection input of multiplexer is controlled by control_in. For control_in = 0, it performs the operation addition, and for control_in = 1, it performs operation subtraction.

The design is shown in Fig. 4.1 and has resources as adders, multiplexers. Here, the issue is the logic is not efficient as both the addition and subtraction are performed at a time, and using the tree of mux at the output, one of the operation outputs is selected. So, the issue is more area, more power and less speed due to cascade stages.

Even the design does not have the better data and control path optimization. So, let us think about logic optimization to have better area, speed and power.

4.2 What Is ALU?

The ALU is an arithmetic logic unit and is used to perform the various arithmetic and logic operations. The design of ALU should be using minimum logic gates, and the goal should be execution of only one operation at a time. The better use of the arithmetic and logic resources can yield into the better ALU design. The following section focuses on the ALU design to have better area and speed (Fig. 4.2).

Strategies for design of the ALU

1. Understand the functional specifications of arithmetic and logic unit.
2. Depending on the number of instructions or operations, find out the opcode required, that is, control input of ALU.

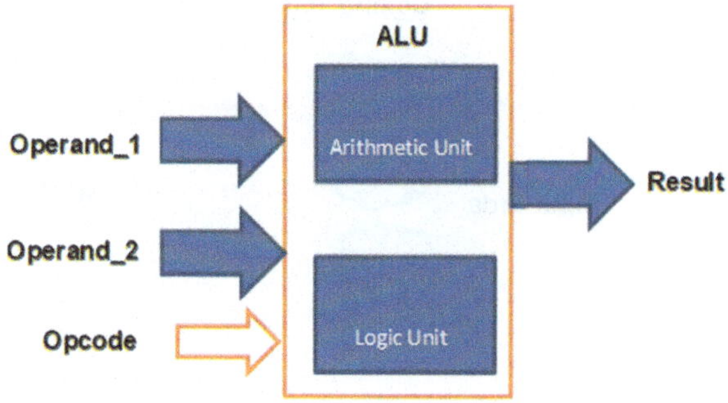

Fig. 4.2 ALU block diagram

3. Design the separate logic for the arithmetic unit.
4. Design the separate logic for the logic unit.
5. Optimize for the area and design the arithmetic and logic unit.
6. Design the ALU to have the least area.

The following section discusses about the design of the arithmetic, logic unit, ALU and their are optimization

4.3 Arithmetic Unit Design

Let us consider the following four operations for which we need to design the arithmetic unit.

- Transfer a_in
- ADD (a_in, b_in)
- SUB (a_in, b_in, 1)
- DECREMENT (a_in).

What we will do is that, we will try to understand about the number of inputs and outputs. For 1-bit arithmetic unit, we need to have 1-bit a_in, b_in, and as four operations are to be performed, we should have 2-bit opcode that is, 2-bit control input. The number of outputs are 2-bit, 1-bit for the result and 1-bit for carry output.

Similarly, for 8-bit arithmetic unit, we need to have 8-bit a_in and b_in, and as four operations are to be performed, we should have 2-bit opcode, that is, 2-bit control input. The number of outputs are 9-bit, 8-bit for the result and 1-bit for carry output.

4.3.1 Resources Required

Let us discuss about the design strategy and resources required. To perform the arithmetic operations, we can use the adders and subtractors as resources in the data

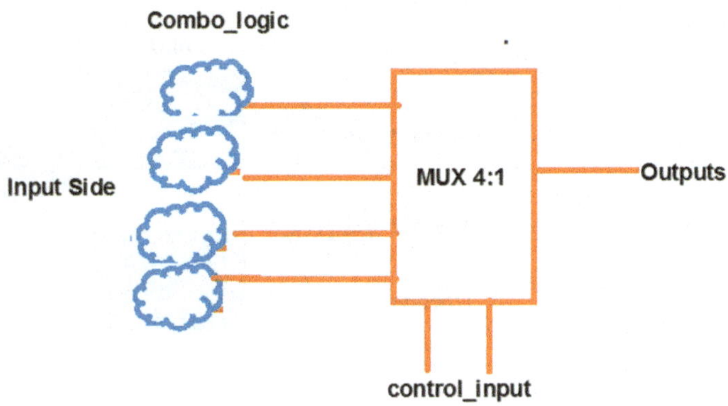

Fig. 4.3 Strategy to design the arithmetic unit

path. To select the result of one of the operations depending on the status of the opcode/control_input, we can use the 4:1 MUX. The strategy is shown in Fig. 4.3.

The main resources are

- Adders
- Subtractors
- Multiplexers.

4.3.2 How to Start Design of ALU?

Now let us document these four operations with their respective opcode that is control input and start analyzing the suitable arithmetic resources. Table 4.2 describes these four operations with the suitable resource.

Table 4.2 Arithmetic instruction description

Control input	Operation	Description	Resource in data path
00	Transfer a_in	Transfer the a_in to output	None
01	ADD (a_in, b_in)	Addition of the a_in, b_in	Adder
10	SUB (a_in, b_in, 1)	Subtraction of a_in, b_in with borrow 1	Full-subtractor or cascade subtractor
11	Decrement (a_in, 1)	Decrement the a_in by 1	Subtractor

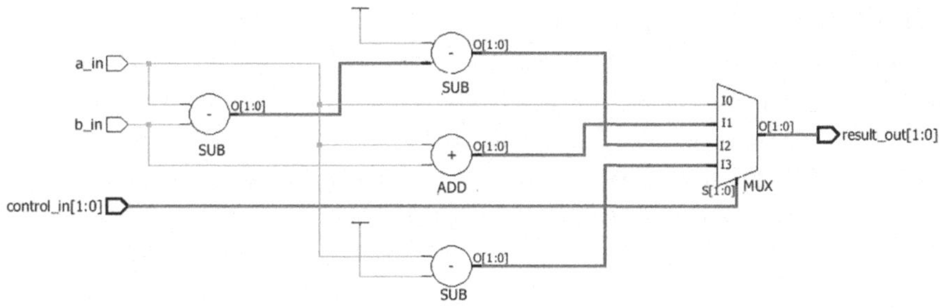

Fig. 4.4 Arithmetic design without resource optimization

4.3.3 How to Design the Logic

From Table 4.2, it is clear that we need to have the arithmetic resources at the inputs of multiplexer. So, now use the 4:1 mux at output and design the logic, use the suitable resource with desired inputs at respective inputs of 4:1 mux and use the control_in[1] and control_in[0] as select inputs of 4:1 mux. The design of arithmetic unit is shown in Fig. 4.4.

The following are the issues in the design of the arithmetic unit:

1. All the operations are executed concurrently, but the mux at the output decides to select the result of only one operation depending on the status of control inputs.
2. Use of the many arithmetic resources without the resource optimization.
3. No proper data and control path optimization.

4.3.4 Exercise 1: Optimization of the Arithmetic Unit

Optimize the design shown in Fig. 4.4: **to get the lesser area using minimum number of resources?**

Solution: In the previous section, we have designed the arithmetic unit and understood about the issues in the design. Now, let us try to optimize the arithmetic unit to have less resources and better data and control path optimization. What we can do is that, we can understand the role of each operation and the resource required and let us finalize the strategy to share the common resources.

The four operations are described in Table 4.3.

As described in Table 4.3, if we use common arithmetic resource as full-adder then for all the operations, one of the input of adder is a_in and another input varies depending on the nature of the instruction. Table 4.4 describes these inputs and outputs depending on the arithmetic operations.

Table 4.3 Arithmetic unit operation entries

Control input	Operation	Description
00	Transfer a_in	a_in + 0 + 0
01	ADD(a_in, b_in)	a_in + b_in + 0
10	SUB(a_in, b_in, 1)	a_in − b_in − 1 = a_in + ~ b_in + 1–1 = a_in + ~ b_in
11	Decrement (a_in, 1)	a_in − 1 = a_in + all 1's

Table 4.4 Inputs and outputs of adder

Control input	X	y
00	a_in	0
01	a_in	b_in
10	a_in	~ b_in
11	a_in	1

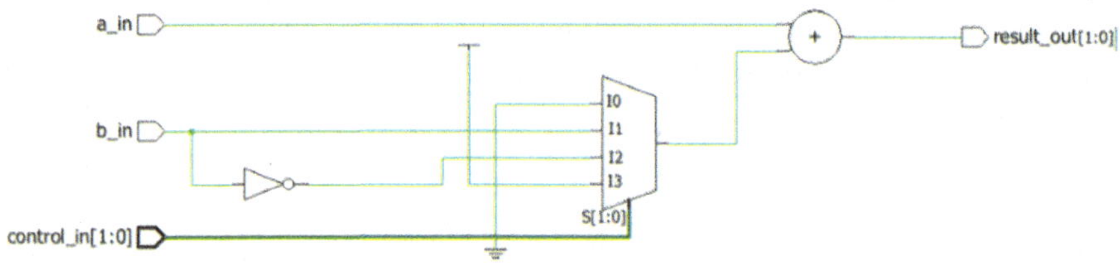

Fig. 4.5 Arithmetic design with resource optimization

So, to improve the area of design, let us use the common resource full-adder at output and let us push the resource 4:1 mux at the input side with the desired input depending on the status of control_in. The arithmetic unit optimized design is shown in Fig. 4.5.

4.3.5 Logic Unit Design

In the previous section, we have discussed about the arithmetic unit design. Even we have discussed about the design optimization using the resource sharing technique. Now let us design the logic unit to perform the OR, NOT, XOR, AND operations. What we need to keep in our mind is that, we should use the minimum resources for optimized logic unit.

Let us consider the following four operations for which we need to design the logic logic.

- OR (a_in, b_in)
- XOR (a_in, b_in)

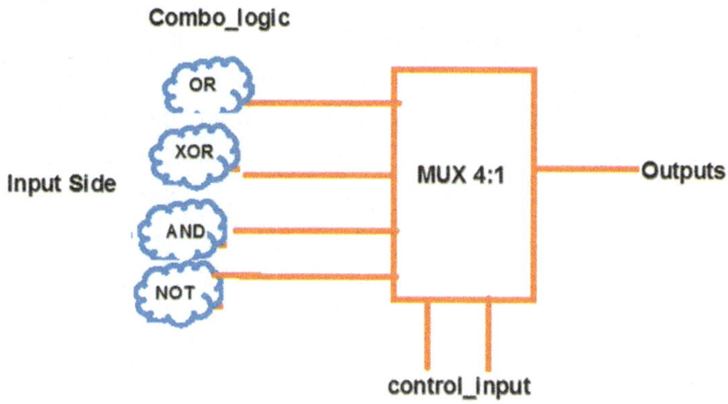

Fig. 4.6 Strategy to design logic unit

- AND (a_in, b_in)
- NOT (a_in).

What we will do is that, we will try to understand about the number of inputs and outputs. For n-bit logic unit, we need to have n-bit a_in, b_in and as four operations are to be performed we should have 2-bit opcode, that is, 2-bit control input. The number of outputs is n-bit.

That is for 1-bit logic unit, we need to have 1-bit a_in and b_in, and as four operations are to be performed, we should have 2-bit opcode, that is, 2-bit control input. The number of outputs is 1-bit.

4.3.6 Resources Required

Let us discuss about the design strategy and resources required. To perform the logic operations, we can use the logic gates as resources in the data path. To select the result of one of the operations depending on the status of the opcode/control_input, we can use the 4:1 MUX. The strategy is shown in Fig. 4.6.

The main resources are

- Logic gates: OR, XOR, AND, NOT
- Multiplexer: 4:1 mux.

4.3.7 How to Design the Logic Unit to have Better Area?

From Table 4.5, it is clear that we need to have the logic gate resources at the inputs of multiplexer. So, now use the 4: 1 mux at output and design the logic, use the suitable

Table 4.5 Logic instructions

Control input	Operation	Description
00	OR (a_in, b_in)	Bit-wise OR of a_in, b_in
01	XOR (a_in, b_in)	Bit-wise XOR of a_in, b_in
10	AND (a_in, b_in)	Bit-wise AND of a_in, b_in
11	NOT (a_in)	Bit-wise NOT of a_in

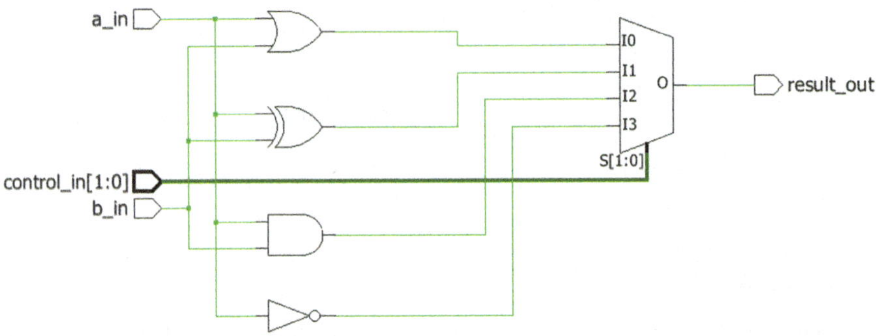

Fig. 4.7 Logic unit design

resource with desired inputs at respective inputs of 4:1 mux and use the control_in[1] and control_in[0] as select inputs of 4:1 mux. The logic unit is shown in the Fig. 4.7.

4.4 ALU Design

Now, let us design the ALU to perform the following eight operations. We can use the arithmetic elements as adders, subtractors and logic gates as the resources.

- Transfer a_in
- ADD (a_in, b_in)
- SUB (a_in, b_in, 1)
- DECREMENT (a_in)
- OR (a_in, b_in)
- XOR (a_in, b_in)
- AND (a_in, b_in)
- NOT (a_in).

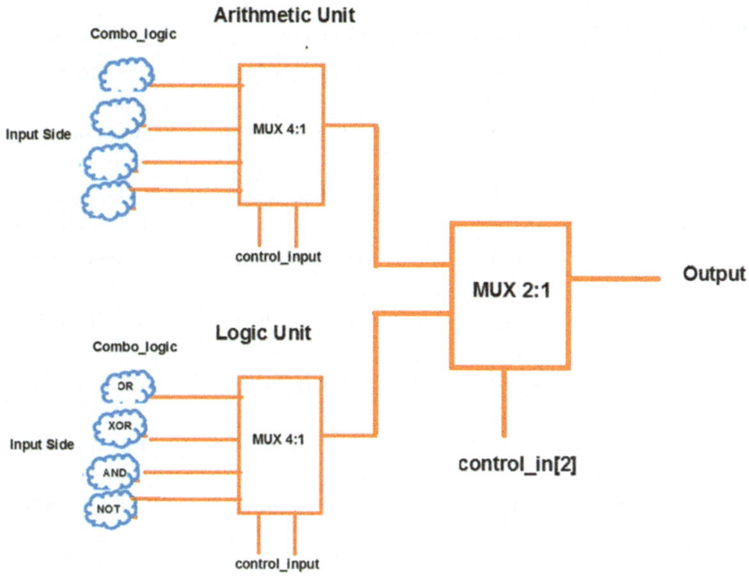

Fig. 4.8 Strategy to design ALU

4.4.1 Resource Requirement and How to Design Efficient ALU?

To perform the arithmetic operations we need to have the adders, subtractors, 2:1 mux and 4:1 multiplexers as resources. To perform the logical operations we can use the OR, AND, XOR, NOT gate and 4:1 mux as resources.

Let us have the separate data path for arithmetic operations and separate control path for the logical operations. As 8 operations need to be performed, let us have control_in as 3-bit opcode. MSB of opcode indicates the operation. That is, control_in[2] = 0 indicates the arithmetic operation and control_in[2] = 1 indicates the logic operation. The strategy is described in Fig. 4.8.

4.4.2 ALU Design to have Better Area

Now to design the efficient ALU, what we can do is that we can tabulate the operations as shown in Table 4.6.

First four operations are arithmetic operations, and next four are logical operations.

As discussed in the previous sections, we can use the mux at output and select either logic or arithmetic operation depending on the MSB of the control_input. The design of ALU which is arithmetic plus logic unit is shown in Fig. 4.9.

Table 4.6 ALU operations

Control input	Operation	Description
000	Transfer a_in	Transfer the a_in to output
001	ADD (a_in, b_in)	Addition of the a_in, b_in
010	SUB (a_in, b_in, 1)	Subtraction of a_in, b_in with borrow 1
011	Decrement (a_in, 1)	Decrement the a_in by 1
100	OR (a_in, b_in)	Bit-wise OR of a_in, b_in
101	XOR (a_in, b_in)	Bit-wise XOR of a_in, b_in
110	AND (a_in, b_in)	Bit-wise AND of a_in, b_in
111	NOT (a_in)	Bit-wise NOT of a_in

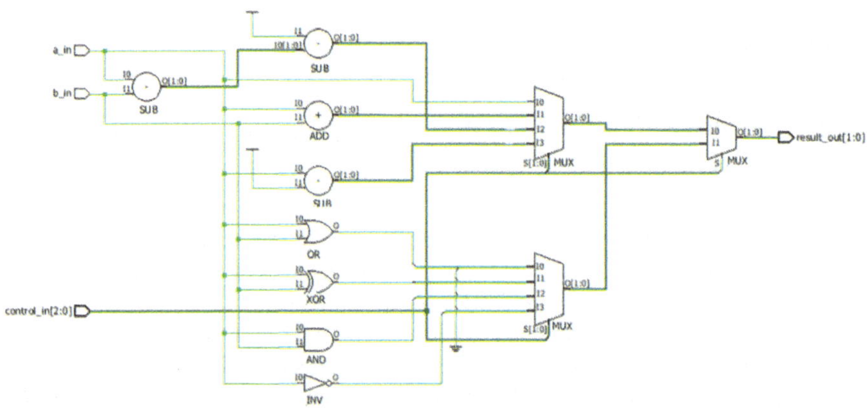

Fig. 4.9 Arithmetic and logic unit design

The following are the issues in the design of the ALU:

1. All the arithmetic and logical operations are executed concurrently, but the mux decides to select the output of only one operation depending on the status of control inputs.
2. Use of the many arithmetic resources without the resource optimization and the multiplexer chain at the output side.
3. No proper data and control path optimization.

Table 4.7 ALU operations with the goal of optimization

Control input	Operation	X	Y
000	Transfer a_in	a_in	0
001	ADD (a_in, b_in)	a_in	b_in
010	SUB (a_in, b_in, 1)	a_in	~ b_in
011	Decrement (a_in, 1)	a_in	1
100	OR (a_in, b_in)	a_in OR b_in	0
101	XOR (a_in, b_in)	a_in XOR b_in	0
110	AND (a_in, b_in)	a_in AND b_in	0
111	NOT (a_in)	NOT of a_in	0

4.4.3 Exercise 2: Optimization of ALU

Optimize the design shown in Fig. 4.9: **to get the lesser area using minimum number of resources?**

Solution: Now as discussed in the arithmetic unit optimization, let us optimize the arithmetic unit using common resource as adder and multiplexer at input side. To avoid the tree of multiplexers at the output side, let us tabulate the operations shown in Table 4.7.

We are using the adder for the arithmetic and logical operations. That means for transfer a_in, the adder will receive a_in from one of the 8:1 mux and logic 0 from other mux. To perform the XOR, the upper mux gives XOR gate output to adder, and another input (logic 0) of adder is from the lower mux. For the logical operation, result_out is single bit, and for arithmetic operation, the result_out is 2-bit. MSB of result_out indicates carry for arithmetic operation. The ALU design is shown in Fig. 4.10.

4.5 Few Important Design Guidelines

If we consider the VLSI design and use of the arithmetic resources to implement the ALU having more than eight instructions, then the following guidelines we can think about!

1. Larger combinational elements in the data path and hence maximum delay. If we have registered inputs and registered output boundaries, then we need to improve the data path by minimizing the combinational area. The techniques like resource sharing are useful in the design.
2. For the fixed and floating point operations, we can think about use of the parallel processing engine to improve the speed and area of the design.
3. To design the logic unit, we can think of use of the arithmetic resources like adders.

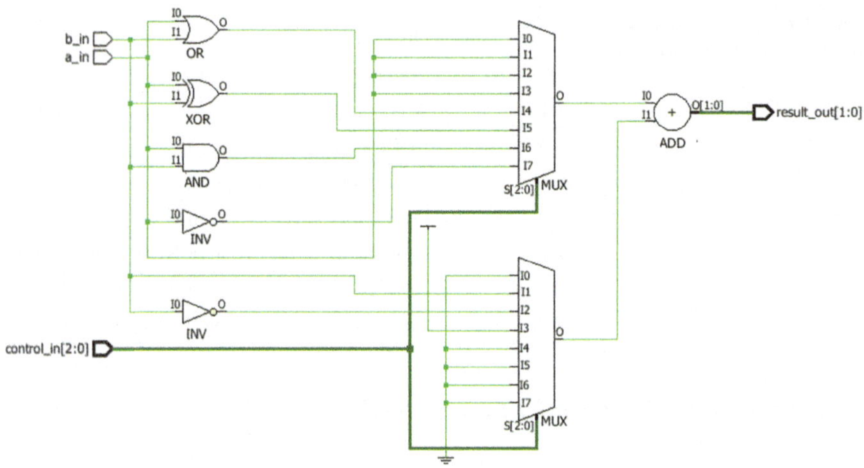

Fig. 4.10 Optimization for the ALU

4. We can design the multipliers and dividers which has low power and lesser area.
5. Instead of the cascade logic, we can think about the parallel logic and we can include the parallelism wherever is required.
6. Have the better strategies to design the data and control path logic.

4.6 Important Takeaways

The following are few of the important points to conclude this chapter.

1. The cascade adders consumes more area, and they have maximum propagation delays.
2. Use the adder as resource to perform both the addition and subtraction.
3. Use the resource sharing technique while designing ALU.
4. Use the common resources at output and try to push the multiplexers at input side.
5. Have the efficient design by using the data and control path optimization.

Chapter 5
Practical Scenarios and the Design Techniques

The understanding of the design scenarios and design techniques used to design the parallel and priority logic are useful during architecture design.

In the previous few chapters, we have discussed about the various combinational logic elements and their use in the design. We have even discussed about the various performance improvement techniques for the design and the exercises to implement the combinational logic. In this chapter, let us discuss about the parallel versus cascade, priority logic and their use in the design.

5.1 Parallel Logic

As the name indicates, the parallel logic has parallel inputs and parallel outputs. But do not get confused with the understanding of the parallel logic. The design techniques used to design the parallel logic should create the design without any hierarchy. We can think about the decoders, demultiplexers and encoders or code converters as parallel logic. The objective of the designer is to have the logic using minimum number of logic gates to have less propagation delay and less area. The following section discusses about the combinational elements such as decoders which are useful to select one of the memories or IO devices in the system design.

5.1.1 Decoder 2 to 4

In most of the interfacing applications, we use the decoders to select one of the memories or IO devices. The goal is to enable the desired chip or logic depending on the address range and to perform the operation of the data transfer.

V. Taraate, *Digital Design Techniques and Exercises*,
https://doi.org/10.1007/978-981-16-5955-3_5

Table 5.1 Truth-table of 2:4 decoder

Enable (en)	s1	s0	y3	y2	y1	y0
1	0	0	0	0	0	1
1	0	1	0	0	1	0
1	1	0	0	1	0	0
1	1	1	1	0	0	0
0	X	X	0	0	0	0

The 2:4 decoder which has active high enable input(en) and active high outputs(only one line is active high at a time during enable condition) is described in Table 5.1. It has select inputs s1 and s0, where s1 is MSB and s0 is LSB. Depending on the status of the select inputs during en = 1, one of the output lines among the y3 to y0 is active high. When en = 0, all output lines y3, y2, y1, y0 are pulled down to logic 0.

How to design the decoding logic?

To design the decoder, let us deduce the product term for every combination of the inputs and outputs as shown in Table 5.1.

For en = 1, s1 = 0, s0 = 0, the product term is $en \cdot \overline{s1} \cdot \overline{s0}$

For en = 1, s1 = 0, s0 = 1, the product term is $en \cdot \overline{s1} \cdot s0$

For en = 1, s1 = 1, s0 = 0, the product term is $en \cdot s1 \cdot \overline{s0}$

For en = 1, s1 = 1, s0 = 1, the product term is $en \cdot s1 \cdot s0$

For en = 0, all the outputs are logic 0 as decoder is disabled. So, the design of 2:4 decoder has Boolean expressions for outputs as shown below.

$$y0 = en \cdot \overline{s1} \cdot \overline{s0}$$

$$y1 = en \cdot \overline{s1} \cdot s0$$

$$y2 = en \cdot s1 \cdot \overline{s0}$$

$$y3 = en \cdot s1 \cdot s0$$

The decoder design using minimum number of AND and NOT gates is shown in Fig. 5.1.

The timing waveform for the various combinations of en, s1, s0 is shown in Fig. 5.2. As shown, one of the output of decoder is high during en = 1 condition. For en = 0, all decoder outputs are logic 0.

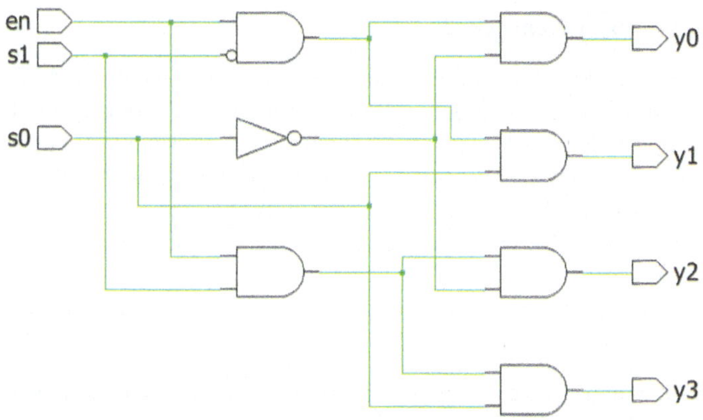

Fig. 5.1 2:4 decoder

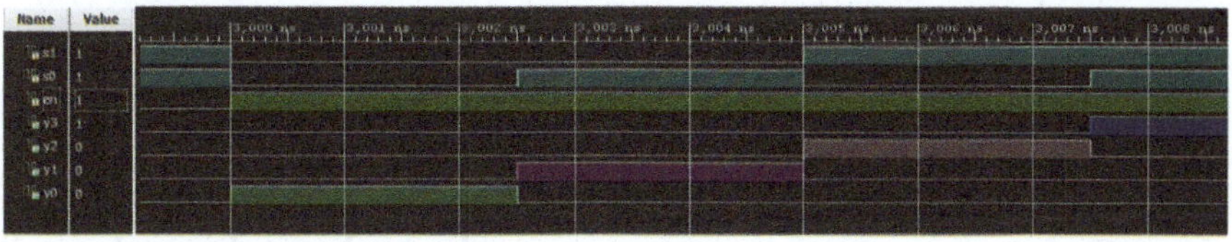

Fig. 5.2 2:4 decoder waveform

5.2 Encoder

As discussed in the previous section, the decoders are used to generate one of the outputs as active at a time during enable condition. The encoders are reverse of the decoder and used to encode the data inputs. For example, for i0 = 1, we need y1 = 0, y0 = 0, and for i3 = 1, we need y1 = 1 and y0 = 1; then, we can think of using 4:2 encoder.

The relationship between the inputs and outputs of encoder is given by $n = \log_2 m$ where m = the number of inputs and n is the number of outputs. For m = 4, the output lines are $n = \log_2 4 = 2$, that is, y1 and y0. Table 5.2 describes the relationship between inputs i3, i2, i1, i0 and outputs y1, y0.

Table 5.2 Truth-table of 4:2 encoder

i3	i2	i1	i0	y1	y0
1	0	0	0	1	1
0	1	0	0	1	0
0	0	1	0	0	1
0	0	0	1	0	0

<u>How to design the 4:2 encoder?</u>

To design the encoder let us use the strategy to input the 16 entries. As having four inputs, and one of the inputs is active high at a time which is our assumption due to that we can consider output for other 12 conditions as x(pronounced as don't care). Now why? As we know that the number of input combinations for the 4 inputs are $2^4 = 16$. But in the truth table of 4:2 encoder, only the outputs are specified for the inputs 1000, 0100, 0010, 0001, so we need to specify outputs as x for remaining input combinations.

So, to deduce the expression for y1 and y0, let us use the 4-varibale K-map (Figs. 5.3 and 5.4).

From the k-map entries, we can get two terms as we have group of eight 1's y1 = i3 + i2, that is, OR of (i3, i2).

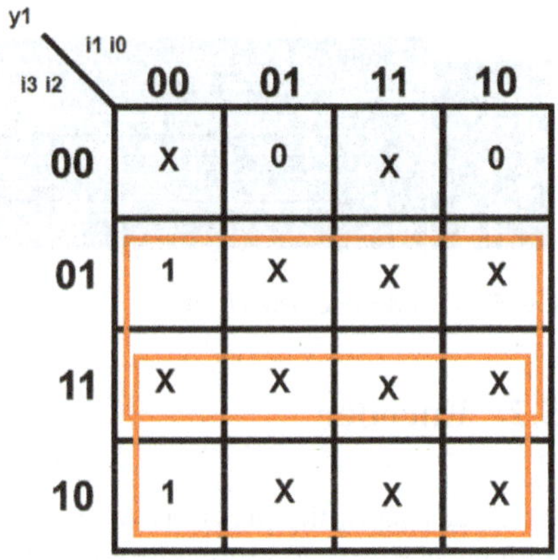

Fig. 5.3 K-map for y1

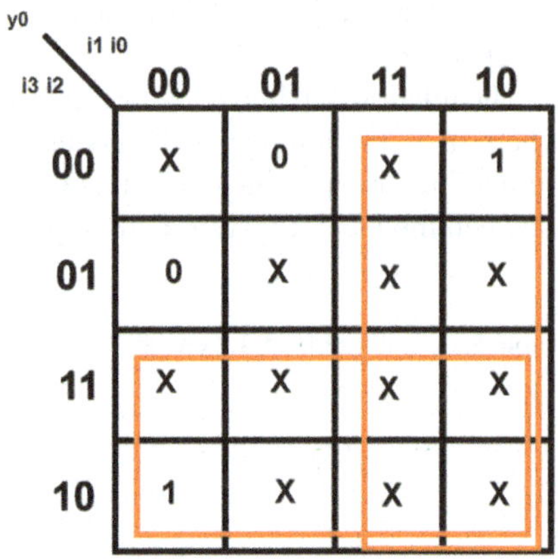

Fig. 5.4 K-map for y0

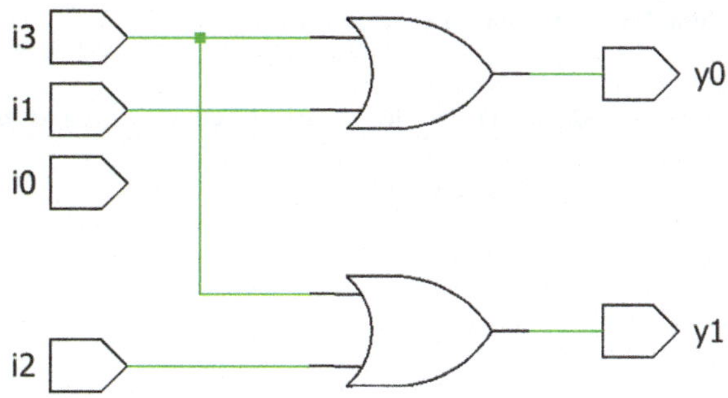

Fig. 5.5 Encoder and the issue of dangling input

From the k-map entries, we can get two terms as we have group of eight 1's y0 = i3 + i1, that is, OR of (i3, i1).

The data input i0 is not connected, and hence, the design has the issues. The logic realized for the 4:2 encoder is shown in Fig. 5.5.

Practical issues in the 4:2 encoder design are listed below:

a. It is assumed that one of the inputs is logic 1, but practically more than one input can be logic 1 at a time.
b. For all the inputs as logic 0, it is not specified what should be output. Output should be invalid.

5.3 Encoder with Invalid Output Detection Logic

As discussed in the above section, we have not made any provision to report the invalid outputs during the design of the encoder if all inputs are logic 0. The encoder should generate invalid output flag as logic 1 when all inputs are logic 0. Table 5.3 describes the 4:2 encoder which can be used in the practical system design.

So, for the invalid output logic, we can have the NOR gate, that is,

$$\text{invalid\_output} = \overline{i3} \,.\, \overline{i2} \,.\, \overline{i1} \,.\, \overline{i0}$$

$$\text{invalid\_output} = \overline{i3 + i2 + i1 + i0}$$

When all the inputs are logic 0, an invalid_output flag is active high (Figs. 5.6 and 5.7).

Thus, y1 = i3 + i2, that is, OR of (i3, i2).

Thus, y0 = i3 + i1, that is, OR of (i3, i1).

Table 5.3 Truth-table of 4:2 practical encoder

i3	i2	i1	i0	y1	y0	Invalid_output
1	0	0	0	1	1	0
0	1	0	0	1	0	0
0	0	1	0	0	1	0
0	0	0	1	0	0	0
0	0	0	0	0	0	1

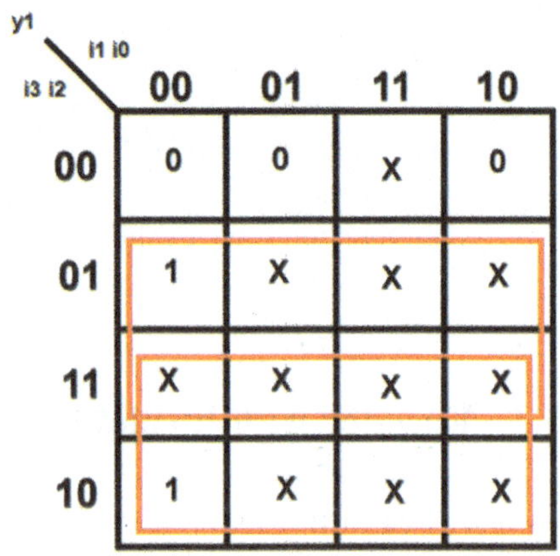

Fig. 5.6 K-map for y1 of encoder

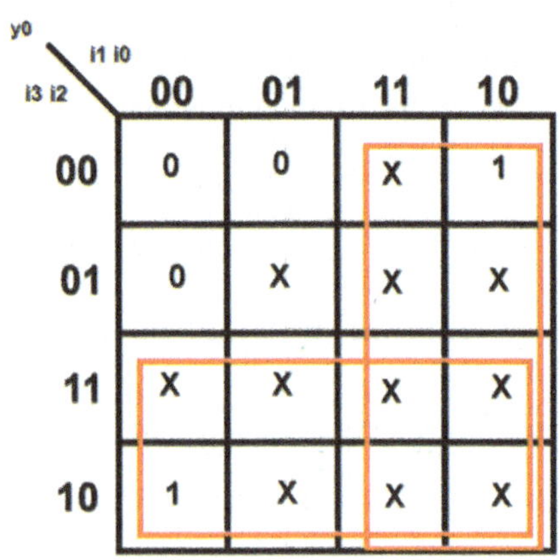

Fig. 5.7 K-map for y0 of encoder

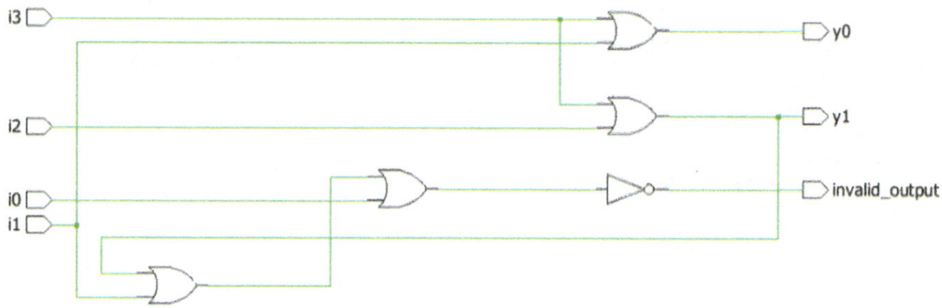

Fig. 5.8 Encoder with the invalid output detection

The logic design of 4:2 encoder having inputs i3, i2, i1, i0 and outputs as y1, y0 and the invalid_output flag are shown in Fig. 5.8.

Still the encoder design shown in Fig. 5.8 has issues, as we have assumed that only one input is 1 at a time, and hence, we need to design the priority encoder design. During the exercises, let us use the understanding of the decoders and encoders and let us try to implement the designs which can be used in few system design applications.

5.4 Exercises

Using the understanding of the decoders, encoders and basic concepts of priority encoder, let us complete the exercises.

5.4.1 Exercise 1: Design of Decoder Having Active-Low Output

Design the 2:4 decoder which has active-low outputs and active-high enable input.

Solution: To design the decoder, let us deduce the product term for every combination of the inputs and outputs shown in Table 5.4.

Table 5.4 Truth-table of 2:4 decoder having active-low output

Enable	s1	s0	y3	y2	y1	y0
1	0	0	1	1	1	0
1	0	1	1	1	0	1
1	1	0	1	0	1	1
1	1	1	0	1	1	1
0	X	X	1	1	1	1

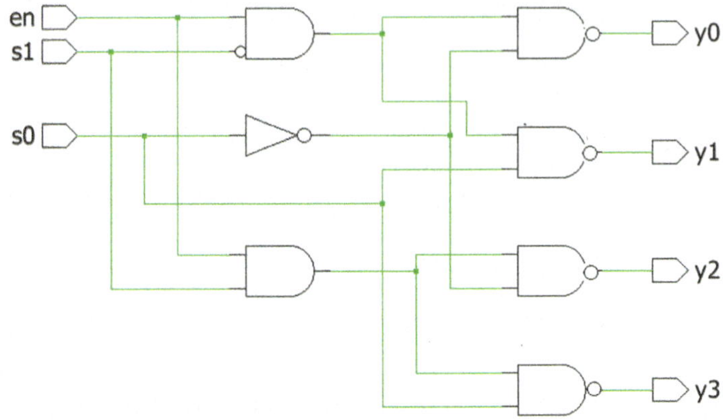

Fig. 5.9 2:4 decoder having active-low outputs

For en = 1, s1 = 0, s0 = 0, the product term is $\overline{\overline{en} \,.\, \overline{s1} \,.\, \overline{s0}}$

For en = 1, s1 = 0, s0 = 1, the product term is $\overline{\overline{en} \,.\, \overline{s1} \,.\, s0}$

For en = 1, s1 = 1, s0 = 0, the product term is $\overline{\overline{en} \,.\, s1 \,.\, \overline{s0}}$

For en = 1, s1 = 1, s0 = 1, the product term is $\overline{\overline{en} \,.\, s1 \,.\, s0}$

For en = 0, all the outputs are logic 1 as decoder is disabled. So, the design of 2:4 decoder has Boolean expressions for outputs

$$y0 = \overline{en.\overline{s1} \,.\, \overline{s0}}$$

$$y1 = \overline{en.\overline{s1} \,.\, s0}$$

$$y2 = \overline{en.s1 \,.\, \overline{s0}}$$

$$y3 = \overline{en.s1 \,.s0}$$

The decoder design using minimum number of NAND and other logic gates is shown in Fig. 5.9.

5.4.2 *Exercise 2: Design the Function Using Decoder*

Implement the Boolean function

1. $f1(s1, s0) = \sum m(1, 2)$
2. $f1(s1, s0) = \sum m(0, 3)$

using the decoder having active-high outputs and active-high enable input. Use minimum logic gates.

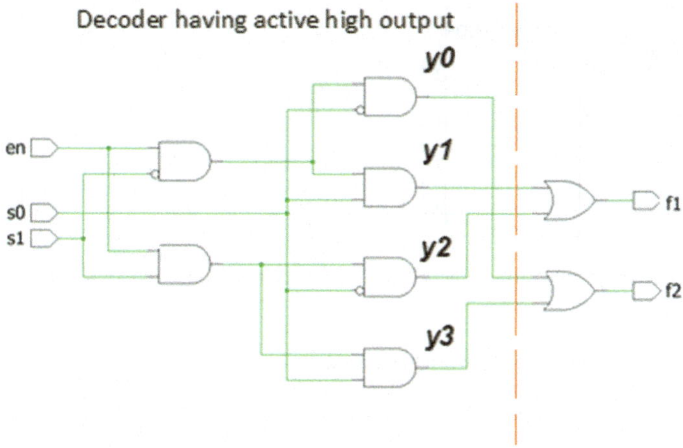

Fig. 5.10 Logic function realization using decoder and logic gates

Solution: Let us use the understanding of the decoder. During en = 1, the decoder has only one output as active high. When en = 0, the decoder all outputs are logic 0 and decoder is disabled.

Now, to realize the function $f1(s1, s0) = \sum m(1, 2)$, *use the OR gate at output as function is SOP. The inputs of OR gate are* y1 *and* y2. *If one of the outputs* y1 *or* y2 *is logic* 1, *then* $f1$ *is logic* 1.

To realize the function $f2(s1, s0) = \sum m(0, 3)$, *use the OR gate at output as function is SOP. The inputs of OR gate are* y0 *and* y3. *If one of the outputs* y0 *or* y3 *is logic* 1, *then* $f2$ *is logic* 1.

The logic realized using the decoder and the OR gates is shown in Fig. 5.10.

5.4.3 Exercise 3: Design Using Decoders

Implement the 2-input XOR and 2-input XNOR using the decoder having active-high outputs and active-high enable input and use minimum logic gates.

Solution: Let us use the understanding of the decoder. During en = 1, the decoder has only one output as active high. When en = 0, the decoder all outputs are logic 0 and decoder is disabled.

Now to realize the logic gate 2-input XOR, let us have the Boolean function $f1(s1, s0) = \sum m(1, 2)$ *use the OR gate at output as function is SOP. The inputs of OR gate are* y1 *and* y2. *If one of the outputs* y1 *or* y2 *is logic* 1, *then* $f1$ *is logic* 1.

To realize the logic gate 2-input XNOR, let us have the Boolean function $f2(s1, s0) = \sum m(0, 3)$ *and use the OR gate at output as function is SOP. The inputs of OR gate are* y0 *and* y3. *If one of the outputs* y0 *or* y3 *is logic* 1, *then* $f2$ *is logic* 1.

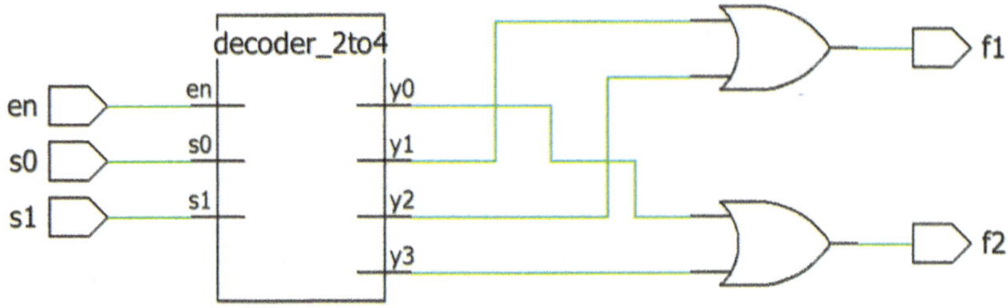

Fig. 5.11 XOR and XNOR logic function realization using decoder and logic gates

Table 5.5 2-input XOR, XNOR logic using decoder

Enable	s1	s0	f1	f2
1	0	0	0	1
1	0	1	1	0
1	1	0	1	0
1	1	1	0	1
0	X	X	0	0

The logic realized using the decoder and the OR gates is shown in Fig. 5.11 (Table 5.5).

5.4.4 Exercise 4: Design Using Decoder and NAND Gates

Implement the 2-input XOR and 2-input XNOR using the decoder having active-low outputs and active-high enable input and use minimum logic gates.

Solution: Let us use the understanding of the decoder. During en = 1, the decoder has only one output as active low. When en = 0, the decoder all outputs are logic 1 and decoder is disabled.

Now to realize the logic gate 2-input XOR, let us have the Boolean function $f1(s1, s0) = \sum m(1, 2)$ *and use the NAND gate at output as function is SOP. The inputs of NAND gate are* $y1$ and $y2$. *If one of the outputs* $y1$ *or* $y2$ *is logic* 0, *then* $f1$ *is logic* 1.

To realize the logic gate 2-input XNOR, let us have the Boolean function $f2(s1, s0) = \sum m(0, 3)$ *and use the NAND gate at output as function is SOP. The inputs of NAND gate are* $y0$ *and* $y3$. *If one of the outputs* $y0$ *or* $y3$ *is logic* 0, *then* $f2$ *is logic* 1.

The logic realized using the decoder and the NAND gates (Bubbled OR) is shown in Fig. 5.12 (Table 5.6.)

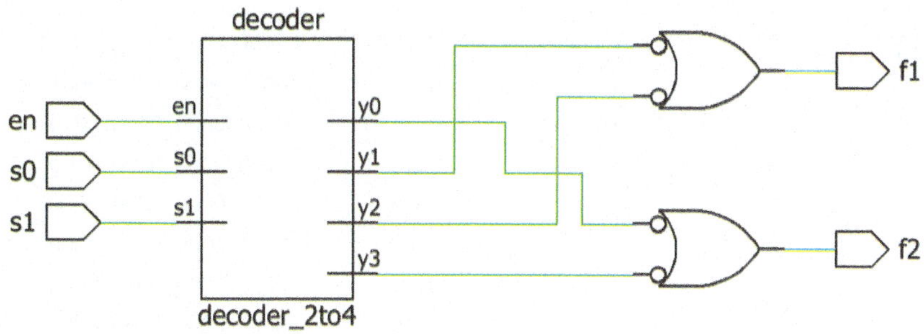

Fig. 5.12 Logic function realization using decoder and bubbled OR gates

Table 5.6 Logic function realization using decoder having active-low outputs

Enable	s1	s0	f1	f2
1	0	0	0	1
1	0	1	1	0
1	1	0	1	0
1	1	1	0	1
0	X	X	0	0

5.4.5 Exercise 5: Design Using Decoders

Design the 4:16 decoder using the minimum number of 2:4 decoder. Consider the decoder has active-high outputs and active-high enable input.

Solution: As 2:4 decoder needs to be used, let us have the truth table (Table 5.7) of the 4:16 decoder having active-high output during en = 1. For en = 0, the decoder all outputs are active low.

Now to get the 16 outputs y0 to y15, let us use the four 2:4 decoders. To select one of the decoders, let us use the 2:4 decoder at input. Table 5.8 gives information about the selection strategy for one of the output decoders.

The 4:16 decoder design using the minimum number of 2:4 decoders is shown in Fig. 5.13. As shown, 4:16 decoder is implemented by using five 2:4 decoders.

5.4.6 Exercise 6: Priority Encoder Design

Design the 4:2 priority encoder using minimum number of logic gates. Consider i3 has highest priority and i0 has lowest priority.

Solution: As it is specified that i3 has highest priority and i0 has lowest priority, let us create the table (Table 5.9) to indicate the relation between the inputs and outputs.

Table 5.7 Truth-table of 4:16 decoder

Enable (en)	s3s2s1s0	Decoder outputs (y15–y0)
1	0000	0000_0000_0000_0001
1	0001	0000_0000_0000_0010
1	0010	0000_0000_0000_0100
1	0011	0000_0000_0000_1000
1	0100	0000_0000_0001_0000
1	0101	0000_0000_0010_0000
1	0110	0000_0000_0100_0000
1	0111	0000_0000_1000_0000
1	1000	0000_0001_0000_0000
1	1001	0000_0010_0000_0000
1	1010	0000_0100_0000_0000
1	1011	0000_1000_0000_0000
1	1100	0001_0000_0000_0000
1	1101	0010_0000_0000_0000
1	1110	0100_0000_0000_0000
1	1111	1000_0000_0000_0000
0	XXXX	0000_0000_0000_0000

Table 5.8 Decoder selection strategy

Enable (en)	s3	s2	Decoder selected
1	0	0	Decoder #1
1	0	1	Decoder #2
1	1	0	Decoder #3
1	1	1	Decoder #4
0	X	X	All outputs are zero

Let us use the invalid_output to indicate the output invalid status when all the inputs i3 to i0 are logic 0.

Now by using the entries specified, let us deduce the expression for the y1 and y0 using 4-varibale K-map.

From the K-map (Fig. 5.14) $y1 = i3 + i2$, that is, OR of (i3, i2).

Thus, for the K-map entries shown in Fig. 5.15: $y0 = i3 + \overline{i2}\,\boldsymbol{i1}$.

For the invalid output logic, we can have the NOR gate, that is,

$$\text{invalid\_output} = \overline{i3}\,.\,\overline{i2}\,.\,\overline{i1}\,.\,\overline{i0}$$

$$\text{invalid\_output} = \overline{i3 + i2 + i1 + i0}$$

When all inputs are logic 0, an invalid_output is active high (Fig. 5.16).

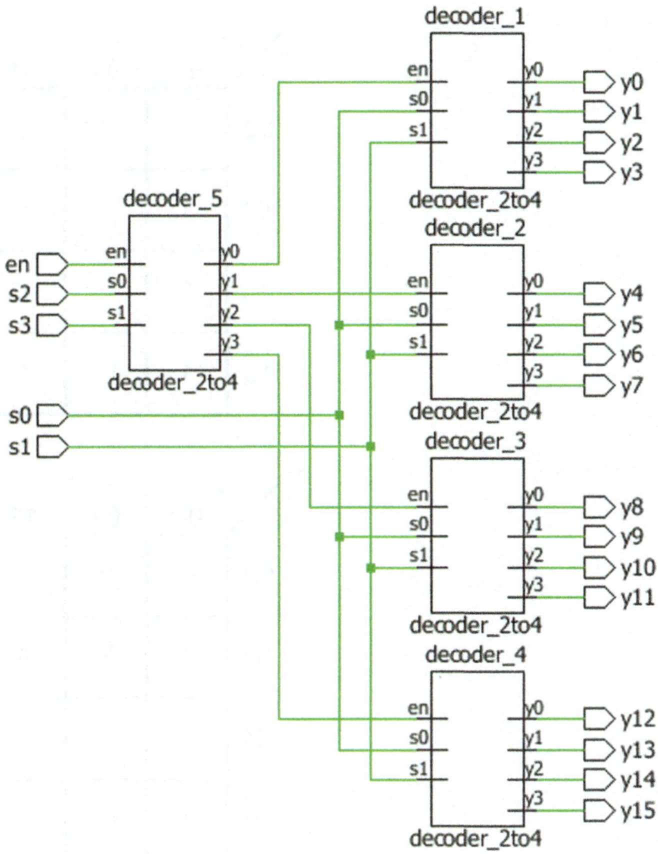

Fig. 5.13 4:16 decoder using 2:4 decoders

Table 5.9 4:2 priority encoder truth-table

i3	i2	i1	i0	y1	y0	Invalid_Output
1	X	X	X	1	1	0
0	1	X	X	1	0	0
0	0	1	X	0	1	0
0	0	0	1	0	0	0
0	0	0	0	0	0	1

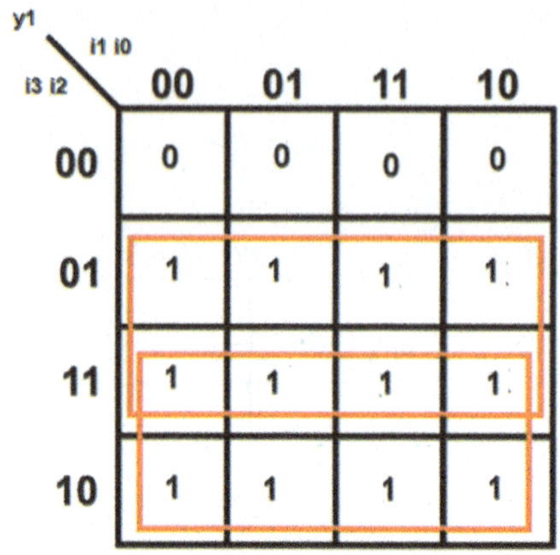

Fig. 5.14 K-map for the y1 of priority encoder

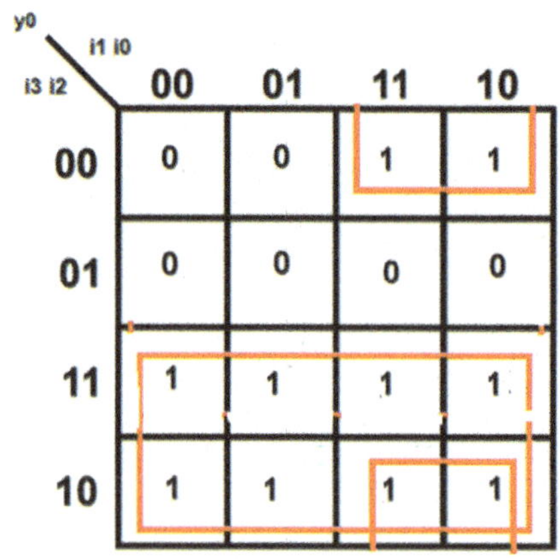

Fig. 5.15 K-map for the y0 of the priority encoder

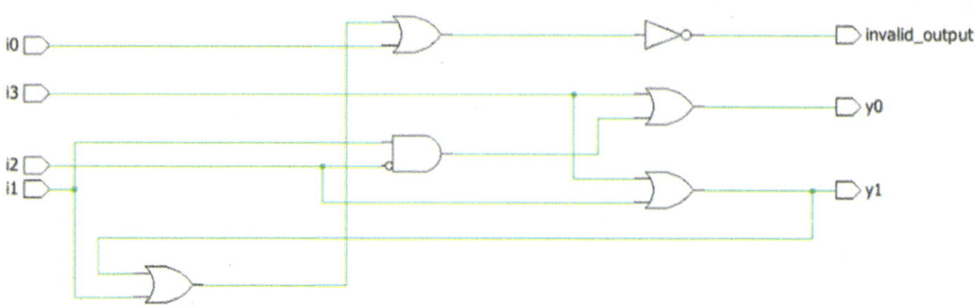

Fig. 5.16 4:2 priority encoder

5.5 Important Takeaways

The following are few of the important points to conclude this chapter.

1. The decoders are used for the selection of one of the memories or one of the IO.
2. For the decoders, one of the outputs is active at a time during enable condition.
3. The relationship between the inputs and outputs of encoder is given by $n = \log_2 m$ where $m =$ number of inputs and n is the number of outputs.
4. The decoders are useful to implement the SOP Boolean functions.
5. The encoders are used to encode the data inputs.
6. The relationship between the inputs and outputs of decoder is given by $m = \log_2 n$ where $m =$ number of inputs and n is number of outputs.
7. The priority encoders are useful in the design of the level-sensitive interrupt controller to identify the highest priority.

Chapter 6
Basics of the Sequential Design

The sequential design uses the present inputs and past outputs to generate an output.

Sequential designs uses the present input and past output to generate an output on the active edge of the clock. The sequential design elements are latches and flip-flops, and they are extensively used in the design. The latch-based designs and flip-flop-based designs and their applications are discussed in this chapter.

6.1 What Is Sequential Logic Design?

In the previous few chapters, we have discussed about the combinational logic in which an output is function of the present input. In the sequential design an output is function of the present input and past output. The sequential design elements are

- Latches
- Flip-flops.

Latches are level-sensitive; flip-flops are edge-triggered. The following section discusses about the roles of these elements in the design.

6.2 Sequential Design Elements

As we know that the latch is level-sensitive, and flip-flop is edge-triggered, these elements and their role are important during the system design and digital design, let us discuss the operation of these elements and timing associated with these elements.

V. Taraate, *Digital Design Techniques and Exercises*,
https://doi.org/10.1007/978-981-16-5955-3_6

6.3 Level Versus Edge-Triggered Logic

As discussed in the previous few chapters the logic gates, combinational logic elements are sensitive to changes in the input. If input changes, an output also changes. In the sequential design, level-sensitive indicates that the output of sequential element changes on the active-level of D latch.

The latches are level-sensitive elements and useful to latch the desired data in some applications. They are used in most of the latch-based design. Consider the design scenario, where the multiplexed buses are used and for demultiplexing we can think of using the multi-bit latch.

The flip-flops, counter or shift registers operate on the active edge of the clock, and they are used in many sequential designs. The active edge can be low to high transition (rising edge or posedge) or high to low transition (falling edge or negedge).

In this section, we will discuss the positive edge-sensitive and negative edge-sensitive flip-flops and their use in the design. For better physical interpretation, it is essential to understand the logic design for the level-sensitive and edge-sensitive elements.

6.4 Latches and Their Use in the Design

As discussed in the previous section, the latches are level-sensitive. Latches are transparent during active level of enable or clock. For example, the output of the positive-level-sensitive D latch is equal to input during the active level that is positive level.

Similarly, the negative-level-sensitive D latch output is equal to the input during negative level.

6.4.1 Positive-Level-Sensitive D Latch

The positive-level-sensitive D latch is transparent during active-high level of enable (EN). The relationship between the inputs of latch and output is shown in Table 6.1.

Table 6.1 Positive-level-sensitive D latch

Enable (sel_in = EN)	Data input (a_in)	Output (y_out)
1	0	0
1	1	1
0	X	Hold the previous output

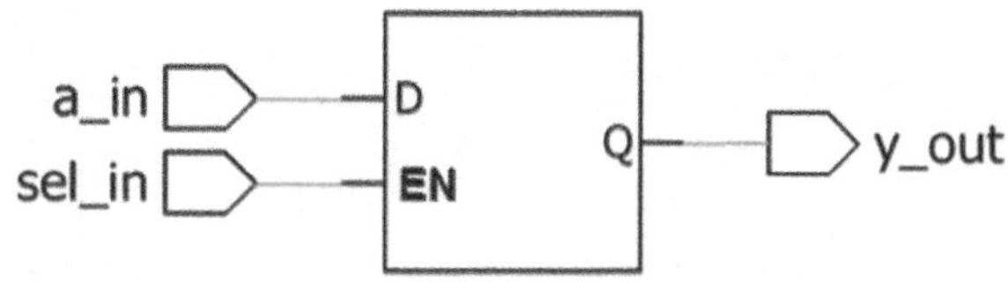

Fig. 6.1 Positive-level-sensitive D latch

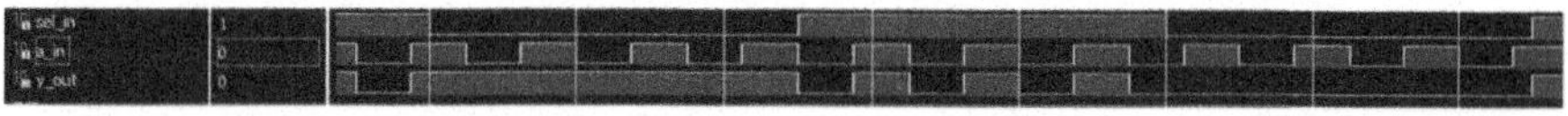

Fig. 6.2 Waveform of the positive-level-sensitive latch

As described in the table for the active high level of the enable that is EN = sel_in output y_out = a_in. For the sel_in = 0, an output y_out is same as the previous output. The D latch holds previous output during inactive level.

The D latch schematic is shown in Fig. 6.1, as shown in the schematic the data input D = a_in, latch enable input EN = sel_in and an output of latch Q = y_out.

The timing sequence of the positive-level-sensitive D latch is shown in Fig. 6.2. As shown, the latch output y_out = a_in for sel_in = 1. The latch holds previous output either 1 or 0 for the sel_in = 0.

6.4.2 Negative-Level-Sensitive D Latch

The negative-level-sensitive D latch is transparent during active-low level of enable (EN). The relationship between the inputs of latch and output is shown in Table 6.2.

As described in the table for the active low level of the enable that is EN = sel_in = 0 output y_out = a_in. For the sel_in = 1, an output y_out is same as the previous output. The D latch holds the previous output during inactive level.

The D latch schematic is shown in Fig. 6.3, as shown in the schematic the data input D = a_in, latch enable input EN = sel_in and output of latch Q = y_out.

The timing sequence of the negative level-sensitive D latch is shown in Fig. 6.4. As shown the latch output y_out = a_in for sel_in = 0. The latch holds the previous output either 1 or 0 for the sel_in = 1.

Table 6.2 Negative-level-sensitive D latch

Enable (sel_in = EN)	Data input (a_in)	Output (y_out)
0	0	0
0	1	1
1	X	Hold the previous output

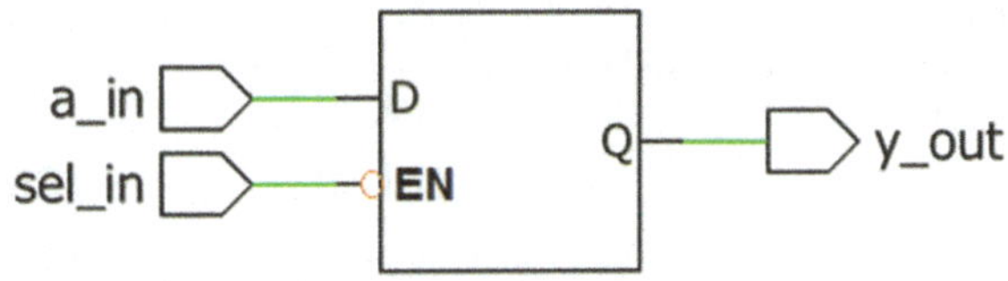

Fig. 6.3 Negative-level-sensitive D latch

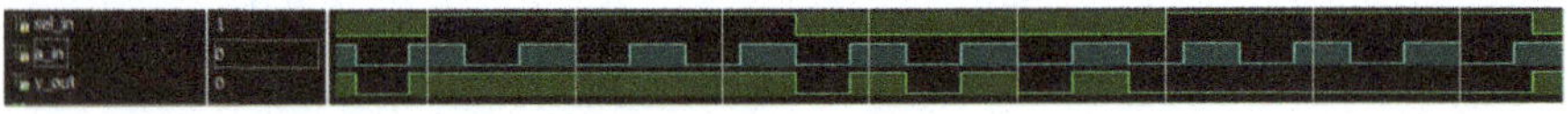

Fig. 6.4 Waveform of the negative-level-sensitive latch

6.5 Edge-Sensitive Elements and Their Role

As we know that most of the time we use the flip-flops as they are sensitive to the active edge of the clock. That means during one clock cycle data is sampled on either positive edge of the clock or on the negative edge of the clock. The main advantage of the flip-flop is that the data is stable for one clock cycle.

We use the D flip-flops either positive or negative edge-sensitive during the design of the counters and shift registers. The logic circuit of the D flip-flops is discussed in this section.

6.5.1 Positive Edge-Sensitive D Flip-Flop

The positive edge means the low to high transition, and it is also called as rising edge. The positive edge-sensitive flip-flop is designed by using two latches in cascade. The data input a_in is given as input to the negative-level-sensitive D latch, and an output of the negative-level-sensitive D latch acts as an input to the positive-level sensitive D latch.

The output of the positive-level-sensitive D latch acts as a flip-flop output. The positive edge-sensitive D flip-flop is shown in Fig. 6.5.

The schematic of the positive edge-sensitive D flip-flop is shown in Fig. 6.6.

The timing sequence of the positive edge-sensitive flip-flop having data input D = a_in, clk = sel_in and an output Q = y_out is shown in Fig. 6.7. As shown the flip-flop samples the data on the active edge of the sel_in that is positive edge of the clock.

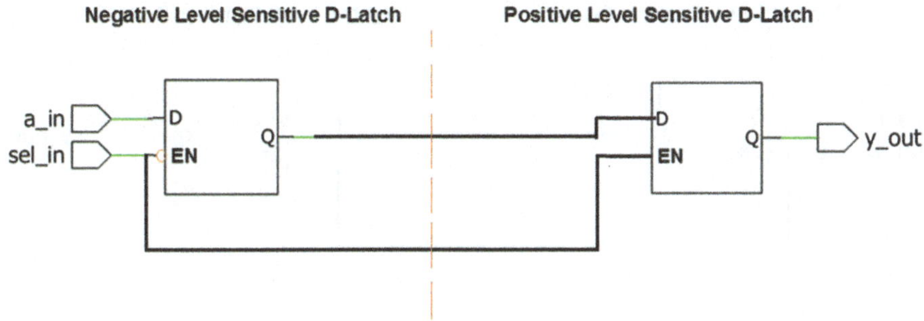

Fig. 6.5 Positive edge-triggered D flip-flop

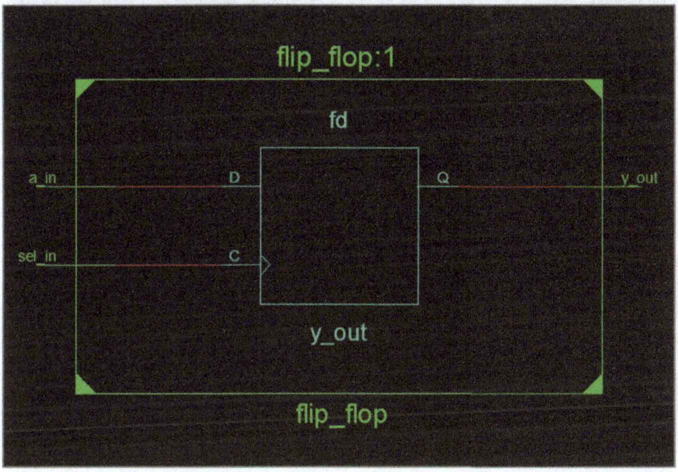

Fig. 6.6 Positive edge-sensitive D flip-flop symbol

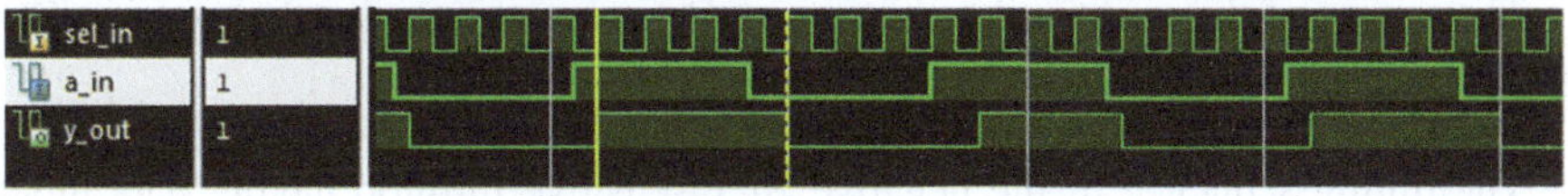

Fig. 6.7 Timing sequence of the positive edge-sensitive D flip-flop

6.5.2 Negative Edge-Sensitive D Flip-Flop

The negative edge means the high to low transition, and it is also called as falling edge. The negative edge-sensitive D flip-flop is designed by using two latches in cascade. The data input a_in is given as input to the positive-level-sensitive D latch, and an output of the positive-level-sensitive D latch acts as input to the negative-level-sensitive D latch.

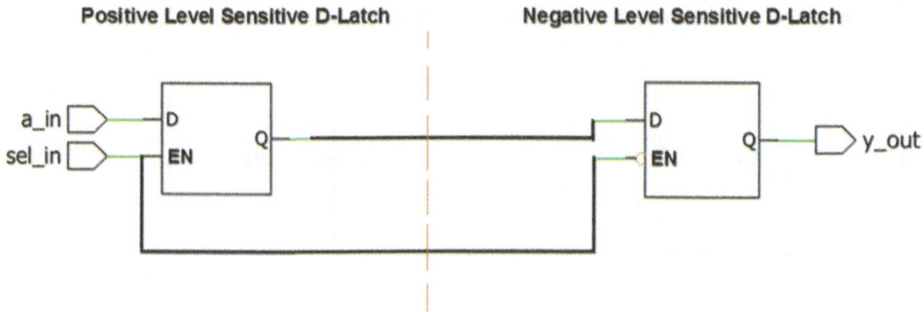

Fig. 6.8 Negative edge-triggered D flip-flop

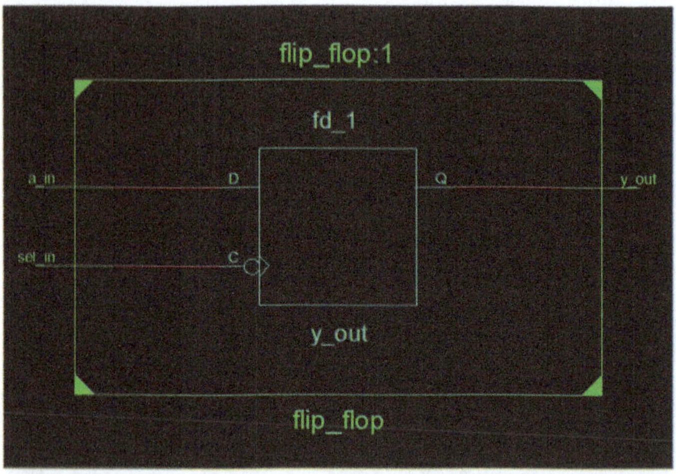

Fig. 6.9 Negative edge-sensitive D flip-flop symbol

The output of the negative-level-sensitive D latch acts as a flip-flop output. The negative edge-sensitive D flip-flop is shown in Fig. 6.8.

The schematic of the negative edge-sensitive D flip-flop is shown in Fig. 6.9.

The timing sequence of the negative edge-sensitive flip-flop having data input D = a_in, clk = sel_in and an output Q = y_out is shown in Fig. 6.10. As shown the flip-flop samples the data on the active edge of the sel_in that is negative edge of the clock.

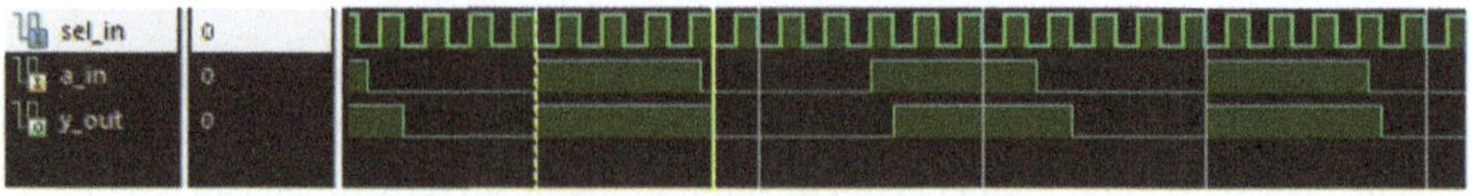

Fig. 6.10 Timing sequence of the negative edge-sensitive D flip-flop

6.6 Applications

Following are few of the applications of latches and flip-flop in the system design.

1. Latches are used in latch-based designs. For example, to demultiplex the buses, we have the latch which acts as transparent during active level and holds the data output during inactive level.
2. The flip-flops either positive edge or negative edge-triggered are used in the design of

   a. Counters
   b. Shift registers
   c. Finite state machine (FSM)
   d. As a registered input element
   e. As a registered output element
   f. Storage registers for the read and write transactions.

The subsequent chapter discusses about the design of these applications with the objective to achieve the desired speed, power and area.

6.6.1 *Applications of the Latches*

Consider the system design scenario which has the processor having 8-bit multiplexed bus. For enable (en) = 1, the bus acts as address bus, and for en = 0, it acts as data bus. To demultiplex the address and data bus we can use the 8-bit latch.

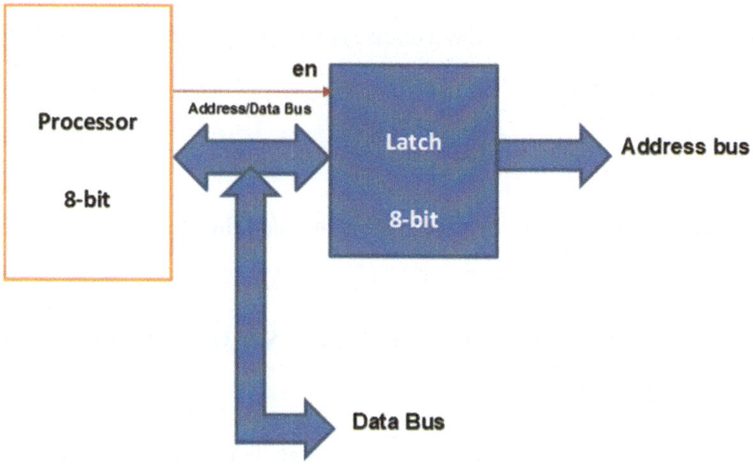

Fig. 6.11 Latch used to demultiplex the bus

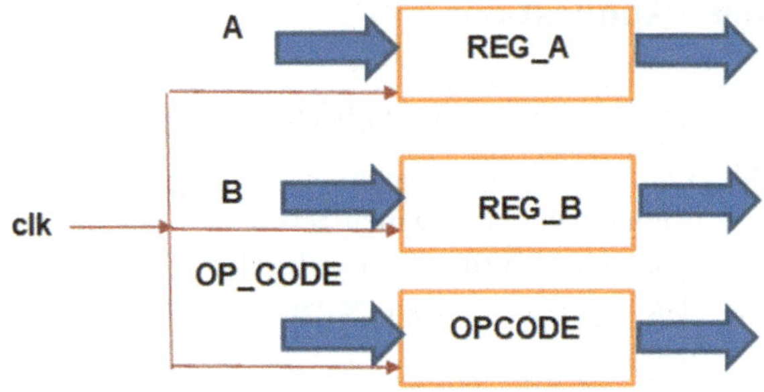

Fig. 6.12 PIPO registers

As shown in Fig. 6.11, the latch is transparent during active-high level of en. So, the address available on the bus is transferred to the output of latch. For en = 0, the latch is disabled and holds the address, and as there is no any connection between the inputs and outputs of latch, the bus acts as a data bus and used to transfer or accept the data.

6.6.2 Applications of the Flip-Flop

The flip-flops are edge-sensitive, and they are used to sample the data input on the active edge of the clock. Consider the registered input for the ALU. To fetch the operands and opcode, the registers are used. Group of flip-flops is register.

As shown in Fig. 6.12, the group of flip-flops (register) are used to sample the operand A, B and op_code of the instruction on the rising edge of the clock. This kind of technique is useful to have the registered input for the ALU or any kind of design.

Let us now discuss about the few exercises on the use of the latches and flip-flops.

6.7 Exercises

Let us complete the exercises by using the latches and flip-flops.

6.7.1 Exercise 1: Design Positive-Level-Sensitive Latch Using Multiplexers

Design the positive-level-sensitive D latch using the minimum number of 2:1 multiplexers only.

Solution: Let us document the positive-level-sensitive D latch (Table 6.3).

Table 6.3 Positive-level-sensitive D latch

Enable (sel_in)	Data input (a_in)	Output (y_out)
1	0	0
1	1	1
0	X	Hold the previous output

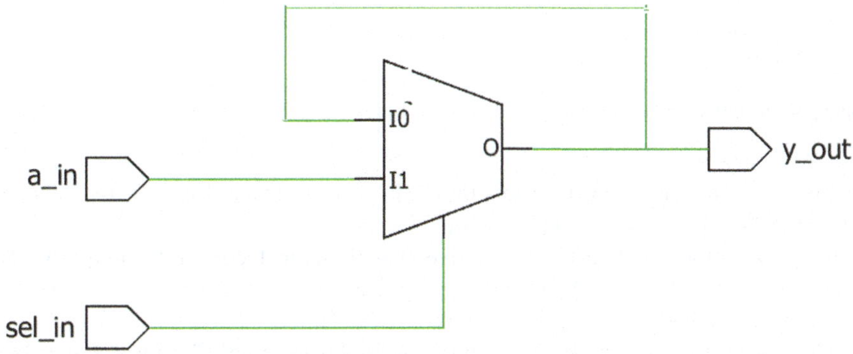

Fig. 6.13 Positive-level-sensitive D latch using 2:1 mux

Now, if we use sel_in as the select input of 2:1 mux, then during positive level of sel_in that is sel_in = 1, the output y_out = a_in.

Now, to get the positive-level-sensitive D latch let us feedback the output y_out to I0 input of the 2:1 mux. During the negative level of the sel_in, the y_out of 2:1 mux is same as the previous output.

The positive-level-sensitive D latch is designed using single 2:1 mux (Fig. 6.13).

6.7.2 Exercise 2: Design Negative-Level-Sensitive Latch Using Multiplexers

Design the negative-level-sensitive D latch using the minimum number of 2:1 multiplexers only.

Solution: Let us document the negative-level-sensitive D latch (Table 6.4).

Table 6.4 Negative-level-sensitive D latch

Enable (sel_in)	Data input (a_in)	Output (y_out)
0	0	0
0	1	1
1	X	Hold the previous output

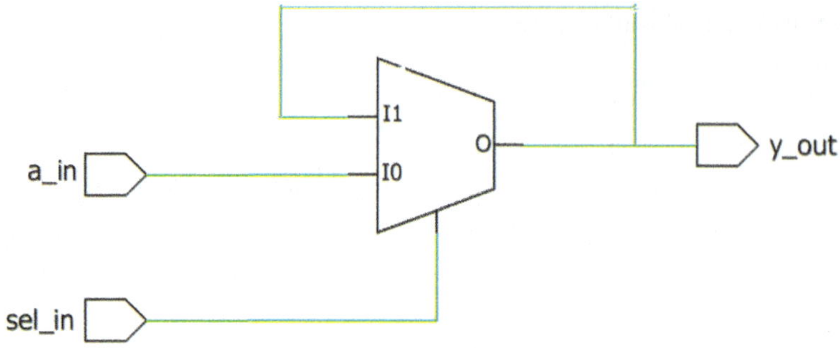

Fig. 6.14 Negative-level-sensitive D latch using 2:1 mux

Now, if we use sel_in as the select input of 2:1 mux, then during negative level of sel_in that is sel_in = 0, the output y_out = a_in.

Now, to get the negative-level-sensitive D latch, let us feedback the output y_out to I1 input of the 2:1 mux. During the positive level of the sel_in, the y_out of 2:1 mux is same as that of the previous output.

The negative-level-sensitive D latch is designed using single 2:1 mux (Fig. 6.14).

6.7.3 Exercise 3: What Is the Functionality of the Following Design?

Can you find what is the functionality of the following logic circuit?

Solution: If you observe Fig. 6.15, then the enable of D latch is controlled by the output of multiplexer. The enable input of latch is NOT of sel_in.

The design samples the a_in during the negative level and holds the previous output during the positive level, and the design is negative-level-sensitive D latch.

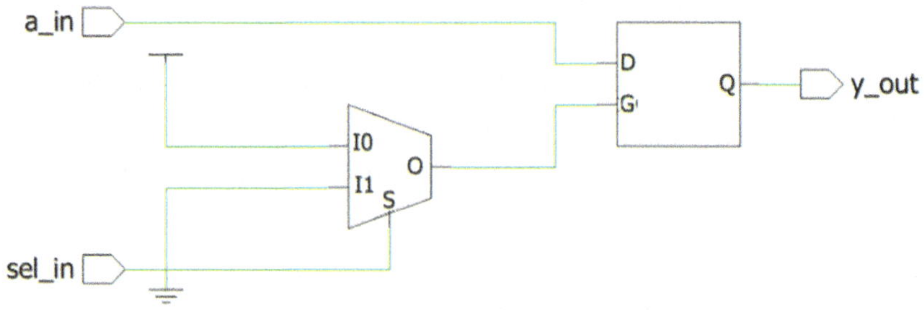

Fig. 6.15 Sequential logic

6.7.4 Exercise 4: Design the Positive Edge-Sensitive Flip-Flop Using Latches

Design the positive edge-sensitive D flip-flop using the minimum number of 2:1 multiplexers only.

Solution: Let us document the positive edge-sensitive D flip-flop (Table 6.5).

For every low to high transition, an output y_out = a_in. So, let us use two level-sensitive latches in cascade. Sample the data input a_in on the negative level of sel_in. The output of the negative-level-sensitive D latch is given as an input to the positive-level-sensitive D latch.

Design the positive and negative-level-sensitive latch using the 2:1 mux as discussed in the exercise 1 and 2. The positive edge-sensitive D flip-flop using the two, 2:1 mux is shown in Fig. 6.16.

Table 6.5 Positive edge-sensitive D flip-flop

clk (sel_in)	Data input (a_in)	Output (y_out)
	0	0
	1	1
	X	Hold the previous output

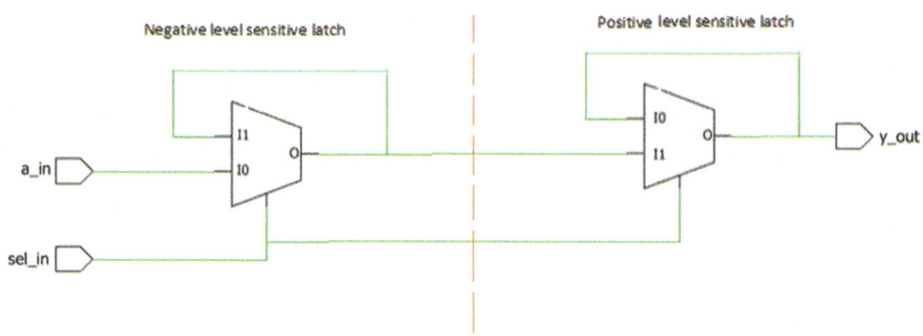

Fig. 6.16 Positive edge-sensitive D flip-flop using 2:1 mux

6.7.5 *Exercise 5: Design the Negative Edge-Sensitive Flip-Flop Using Latches*

Design the negative edge-sensitive D flip-flop using the minimum number of 2:1 multiplexers only.

Solution: Let us document the negative edge-sensitive D flip-flop (Table 6.6).

For every high to low transition, an output y_out = a_in. So, let us use two level-sensitive latches in cascade. Sample the data input a_in on the positive level of sel_in. The output of the positive-level-sensitive latch is given as an input to the negative-level-sensitive D latch.

Design the positive and negative-level-sensitive latch using the 2:1 mux as discussed in the exercise 1 and 2. The negative edge-sensitive D flip-flop using the two 2:1 mux is shown in Fig. 6.17.

Table 6.6 Negative edge-sensitive D flip-flop

clk (sel_in)	Data input (a_in)	Output (y_out)
	0	0
	1	1
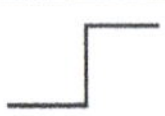	X	Hold the previous output

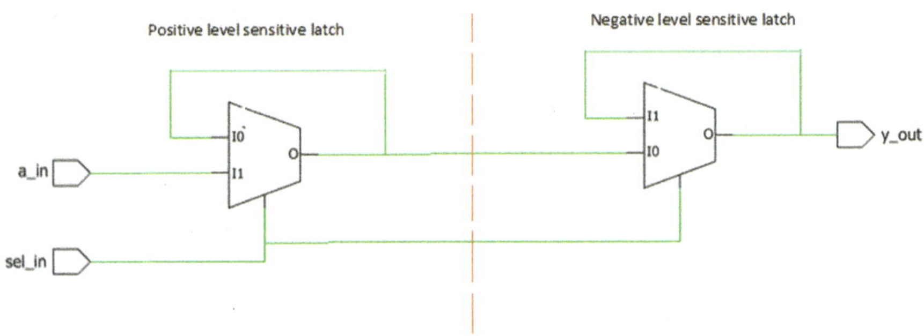

Fig. 6.17 Negative edge-sensitive D flip-flop using 2:1 mux

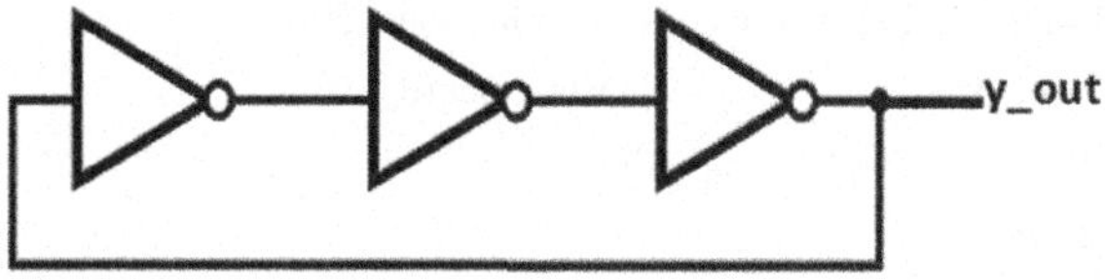

Fig. 6.18 Ring oscillator

6.7.6 Exercise 6: What Is the Operating Frequency of the Following Circuit?

Consider the delay of each NOT gate is 0.5 nanosecond. Find the operating frequency at y_out for the ring oscillator (Fig. 6.18)?

Solution: As three NOT gates are connected in cascade, the output y_out will toggle for delay of the 3 * tpd = 3 * 0.5 nanosecond.

The clock frequency at y_out of the ring oscillator is

$$
\begin{aligned}
f_{\max} &= \frac{1}{T} \\
&= \frac{1}{2 * 3 * 0.5 \text{ nano-second}} \\
&= \frac{1}{3 \text{ nano-second}} \\
&= 333.33 \text{ MHz}
\end{aligned}
$$

6.7.7 Exercise 7: The Asynchronous Clear

Sketch the D flip-flop which has asynchronous active-low reset. Document the truth-table for the flip-flop.

Solution: Asynchronous reset or asynchronous clear is used to initialize the flip-flop, and the output is forced to logic 0 when reset_n = 0 irrespective of the active edge of the clock.

Table 6.7 gives information about the inputs and outputs of the flip-flop.

Table 6.7 Flip-flop having asynchronous reset

reset_n	Data input (d_in)	Output (y_out)
1	0	0
1	1	1
0	X	0

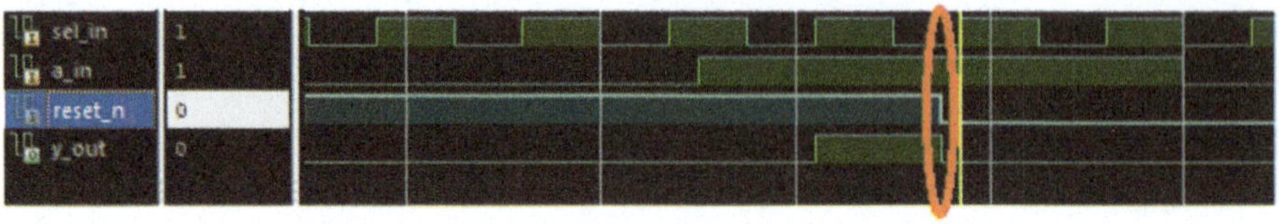

Fig. 6.19 Timing sequence for the asynchronous input

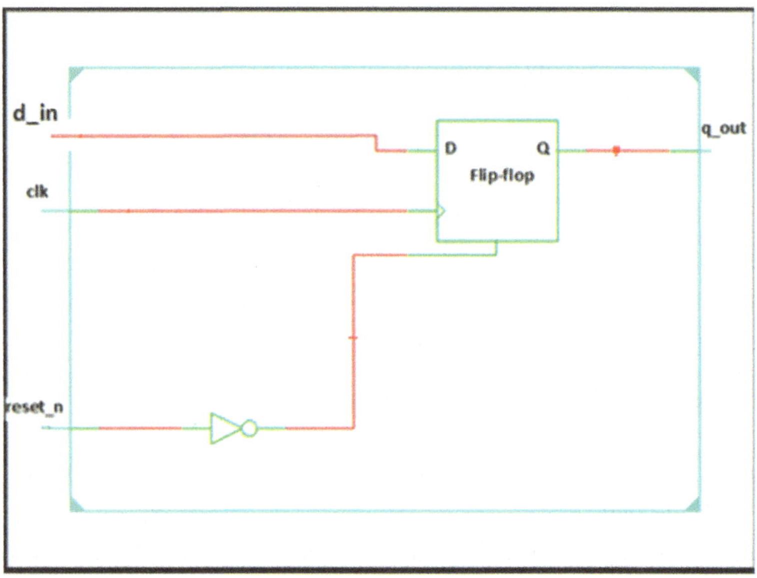

The timing sequence is shown in Fig. 6.19. As shown, the output y_out is forced to logic 0 irrespective of the rising edge of the clock (sel_in) when reset_n = 0.

6.7.8 *Exercise 8: The Synchronous Clear*

Sketch the D flip-flop which has synchronous active-low reset. Document the truth-table for the flip-flop, and sketch the timing sequence.

Table 6.8 Flip-flop having synchronous reset

reset_n	clk	Data input (d_in)	Output (y_out)
1	↑	0	0
1	↑	1	1
0	↑	X	0

Solution: Synchronous reset or synchronous clear is used to initialize the flip-flop, and the output is forced to logic 0 when reset_n = 0 on the active edge of the clock.

Table 6.8 gives information about the inputs and outputs of the flip-flop (Fig. 6.20).

The timing sequence is shown in Fig. 6.21. As shown, the output y_out is forced to logic 0 on the rising edge of the clock (sel_in) when reset_n = 0. If reset_n = 0

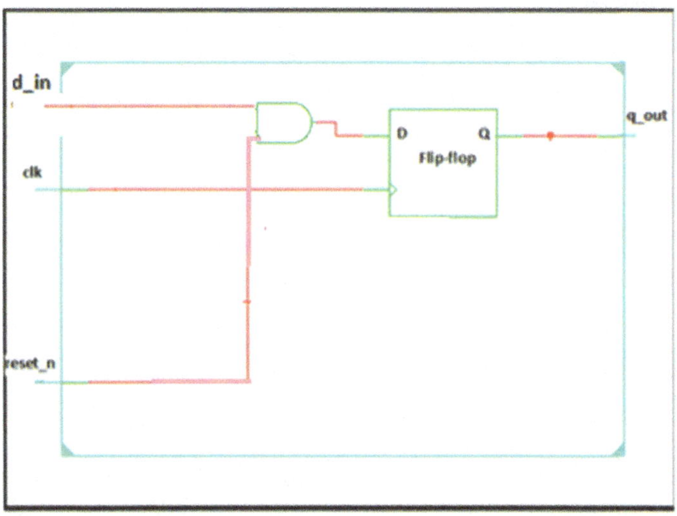

Fig. 6.20 D flip-flop having synchronous reset

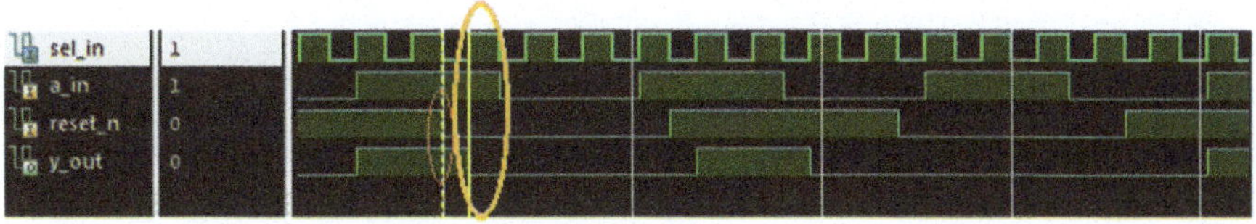

Fig. 6.21 Timing sequence for the synchronous input

prior to the active edge of the clock, then the output is not forced to logic 0 as the type of the reset is synchronous reset.

6.8 Important Takeaways

Following are few of the important points to conclude this chapter.

1. The sequential design elements are latches and flip-flops.
2. In the sequential design, an output is function of the present inputs and past outputs.
3. The D latch is level-sensitive.
4. The D flip-flop is edge-sensitive.
5. The latches are transparent during active level of the enable input.
6. The flip-flop samples data input on the active edge of the clock.
7. To design the counters and shift registers, we will use the flip-flops as they are edge-sensitive.

Chapter 7
Sequential Design Techniques

The design techniques to implement the sequential logic are useful in various system design applications.

In the previous few chapters, we have discussed about the combinational and sequential elements. Let us use the logic gates and flip-flops to design the sequential logic. The chapter discusses about the various techniques useful to implement the sequential designs. The goal is to have the sequential design which has lesser area, maximum speed and low power.

We have two types of sequential designs as follows:

- Synchronous design
- Asynchronous design.

7.1 Synchronous Design

In the synchronous design the clock is common to all the flip-flops. That is clock is derived from the common clock source PLL. PLL is phase-locked loop and used to generate the clock. Examples are counters, shift registers where all the elements are using the common clock. The synchronous design having common clock source is shown in Fig. 7.1.

7.2 Asynchronous Design

In the asynchronous design the clock of the various sequential elements is driven by the different clock source. As shown in Fig. 7.2, the LSB flip-flop receives the clock from the PLL, and the MSB flip-flop receives the clock as one of the outputs of the LSB flip-flop.

V. Taraate, *Digital Design Techniques and Exercises*,
https://doi.org/10.1007/978-981-16-5955-3_7

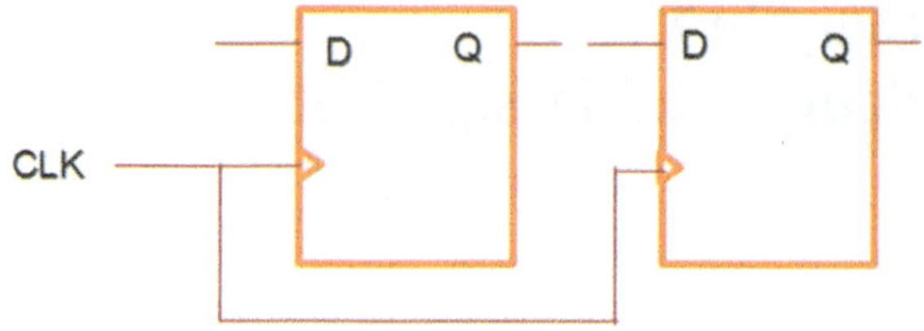

Fig. 7.1 Synchronous design

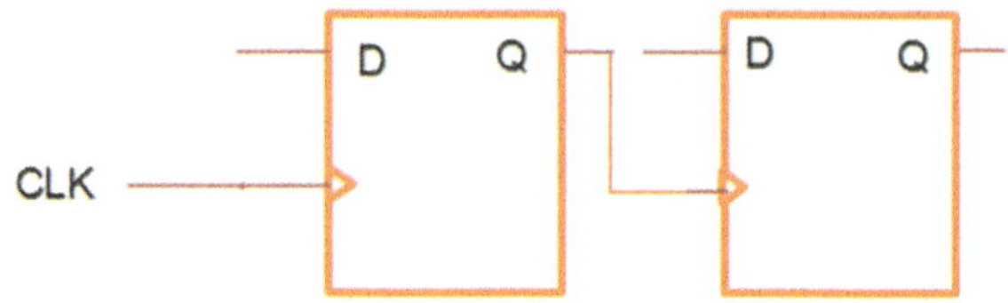

Fig. 7.2 Asynchronous design

7.3 Why to Use Synchronous Design?

As discussed earlier in the asynchronous design, the clock of the MSB flip-flop is derived from the output of the previous stage. Depending on the operation requirement, the clock of the flip-flop is derived, and the LSB flip-flop receives the clock as PLL output.

Now, let us try to understand what is the issue in the asynchronous design? In the asynchronous design, the overall delay to generate the output is due to propagation of the clock. For the n stages, the delay to get the output from the MSB flip-flop is n * tpff where tpff is flip-flop propagation delay or clock to q delay. The delays and advanced concepts like timing parameters of flip-flop are discussed in the next few chapters.

For the asynchronous design shown in Fig. 7.3, the number of stages is $n = 4$, and hence, the maximum propagation delay of the design is 4 * tpff. If tpff is 1 ns (pronounced as 1 ns), then the delay to get the output from MSB of flip-flop is 4 ns.

As discussed earlier, the asynchronous designs have the more delay, and they are slower as compared to the synchronous design. So, in most of the applications and in the system design, we need to have the high speed, and hence, the asynchronous design use is ruled out. Even asynchronous clocking has the issues and hence treated as poor clocking scheme.

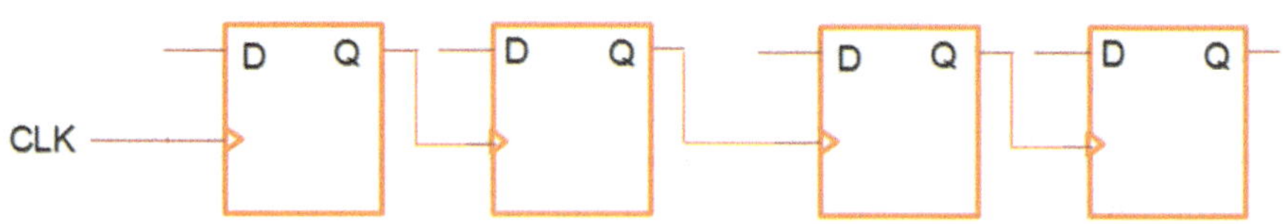

Fig. 7.3 Asynchronous design having D flip-flop stages

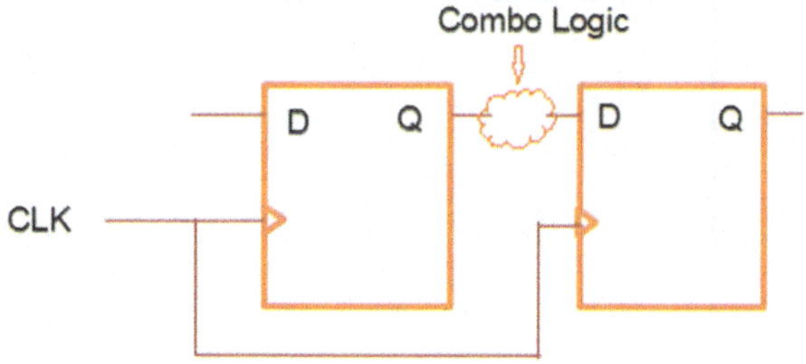

Fig. 7.4 Synchronous design (reg-to-reg path)

Figure 7.4 shows the synchronous design where clock of both the flip-flops is derived from the common clock source. In between the flip-flops, the combinational logic is used, and most of the time we call this as reg-to-reg path. Here, the maximum frequency of the design is dependent on the timing parameters of the flip-flops that is setup time (tsu), hold time (th), clock to *q* delay (tctoq or tpff) and the combinational logic delay.

The maximum frequency calculation and the timing parameters of the flip-flops are discussed in the subsequent chapter. The following section discusses about the design techniques to implement the sequential design.

7.3.1 Which Elements We Should Use During Design?

We can use the sequential element as either positive edge-triggered or negative edge-triggered flip-flops. The flip-flop samples the data on the active edge of the clock and used in the sequential design. To implement the counters, shift registers or finite state machines, we can use the other suitable elements such as logic gates or combinational elements depending on the design requirement.

For example, consider the divide by two counter that is toggle flip-flop designed by using D flip-flop and additional combinational elements (Fig. 7.5). The flip-flop is rising edge-triggered and has the asynchronous active-low reset.

Reset path: The NOT gate is used in the reset path and the flip-flop has asynchronous active-low reset. Asynchronous reset or clear means the flip-flop output is initialized to logic 0 when reset input is active low (logic 0) irrespective of the active edge of the clock.

DATA path: The NOT gate is used in the reset data path as the requirement is to get the toggle output on the rising edge of the clock. The NOT gate complements the output of flip-flop and can be optimized using the complement of Q output of the flip-flop.

Clock path: The design has clock, which is generated from the PLL, and it is named as clk. For the low power designs, the clock path has the additional logic and discussed in Chap. 11. The clock gating cells are useful to minimize the power dissipation.

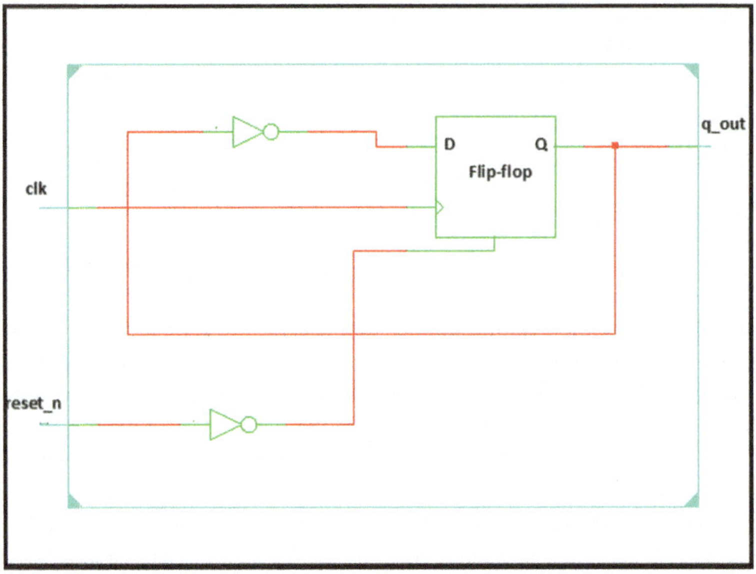

Fig. 7.5 Toggle D flip-flop

Maximum frequency for the design: Each sequential design operates on the specific frequencies, and the maximum frequency of the design is restricted due to the tsu, th, tctoq and the combinational delay. For the divide by 2 counter that is toggle flip-flop, the maximum frequency is equal to

$$\text{fmax} = \frac{1}{\text{tpff} + \text{tcombo} + \text{tsu}}$$

where tpff = flip-flop propagation delay which is also called as clock to q delay
tcombo = tinv = That is combinational delay which is NOT gate delay
tsu = setup time of flip-flop. That is minimum amount of time for which data should be stable at D input of flip-flop prior to rising edge of the clock.

The maximum frequency calculations and the details of the timing paths and parameters are discussed in the next subsequent chapters.

7.4 D Flip-Flop and Use in the Design

As D flip-flop has the single input, the D flip-flops are popular to design the sequential logic. It is easy to control the single D input as compared to the JK or SR flip-flops two inputs. Due to the less logic in the data path, the area requirement for the sequential design using the D flip-flops is lesser as compared to SR or JK flip-flop.

Table 7.1 State table of D flip-flop

Present state (Q)	Next state (Q^+)
0	0
0	1
1	0
1	1

Table 7.2 Excitation input of the D flip-flop

Present state (Q)	Next state (Q^+)	Excitation input (D)
0	0	0
0	1	1
1	0	0
1	1	1

What we need to understand about the D flip-flop?

For the sequential design using the D flip-flop, we need to have the good understanding about the excitation inputs and excitation table. So, let us try to discuss!

Consider the present state of the D flip-flop is Q and the next state of the D flip-flop is Q^+ (Table 7.1). The excitation table gives information about what should be the data input of the D flip-flop to get the next state output on the active edge of the clock. The excitation table of the D flip-flop is documented (Table 7.2).

As shown in Table 7.2, the D input of the flip-flop is equal to the next state Q^+. That means, if the present state $Q = 0$ and the next state $Q^+ = 1$, then to get the next state as output of the flip-flop, the data input of flip-flop $D = 1$. That is $D = Q^+$. During the sequential circuit designs and the FSM designs, we use the excitation table to deduce the combinational logic required at the data input of D flip-flop.

Now let us consider the design scenario.

Consider the design requirement to get the toggle output using the single D flip-flop. What we need to do is that let us document the entries in the excitation table and let us deduce the logic.

As shown in Table 7.3, the excitation input $D = Q^+$ which is the complement of the present state Q. So, $D = \overline{Q}$. So, the toggle flip-flop is designed by using the single D flip-flop and NOT gate. If the flip-flop uses the asynchronous active-low reset, then in the reset path the clear input of the flip-flop is controlled by NOT gate.

The design is shown in Fig. 7.5.

Table 7.3 Excitation table of the D flip-flop

Present State (Q)	Next State (Q^+)	Excitation Input (D)
0	1	1
1	0	0

7.5 Design for the given specifications

Now, let us use the design specifications to implement the sequential design.

1. The design should use the positive edge-sensitive flip-flops.
2. The output of design should toggle on the rising edge of the clock.
3. The design should have active-low asynchronous clear input.
4. The design uses the asynchronous active-high enable that is for data_in = 1 the flip-flop toggles.

Now, let us use the design specifications and try to design the sequential logic.

The design requirement is to have the toggle output on every rising edge of the clock. We have two states that is flip-flop output logic 1 is state s1 and flip-flop output logic 0 is state s0. So, as 2-states, we should have the single flip-flop. The relationship between the number of states and flip-flop is that number of states is equal to 2 to the power of number of flip-flops.

Hence, to get the toggle output, we need to have single D flip-flop. So, we will use the single positive edge-sensitive flip-flop. The design specifications are tabulated in the truth table.

As shown in Table 7.4, the excitation input $D = Q^+$ which is the complement of the present state Q for enable input data_in = 1 so, $D = \overline{Q}$. So, the toggle flip-flop is designed by using the single D flip-flop and NOT gate. The enable input is incorporated in the design which holds the previous state of the flip-flop during data_in = 0 condition. If the flip-flop has the asynchronous active-low reset, then in the reset path, the clear input of the flip-flop is controlled by NOT gate.

The design is shown in Fig. 7.6.

Table 7.4 Excitation table with the enable input

Enable (data_in)	Present state (Q)	Next state (Q^+)	Excitation input (D)
1	0	1	1
1	1	0	0
0	0	0	0
0	1	1	1

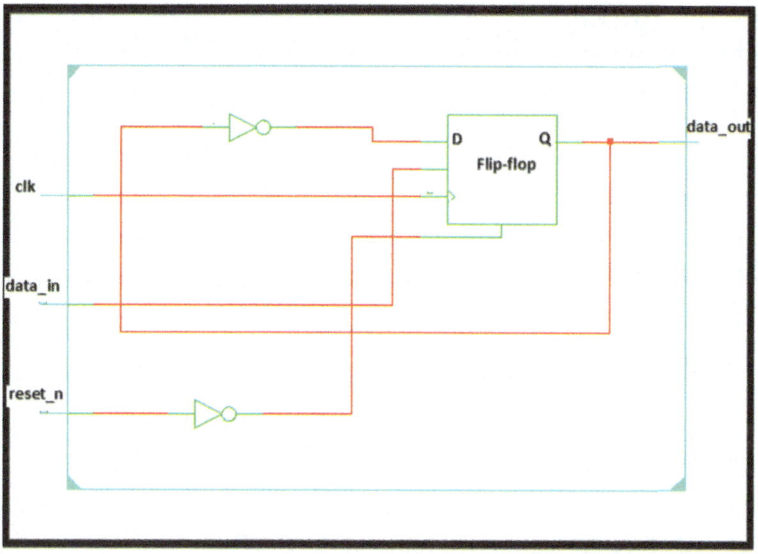

Fig. 7.6 Toggle flip-flop with enable as input

7.6 Design of the Synchronous Counters

Now, let us try to use the design foundation discussed in the previous sections to design the synchronous modulo-4 or MOD-4 binary up-counter. As name itself indicates, the counter has the four states. The four states are so, s1, s2, s3 and for the binary up-counter design use the following steps.

1. **Find the number of states to get the counter**

Number of states = 4.

2. **Find number of flip-flops**

Number of flip-flops $= n = \log_2 4 = 2$. We will use the positive edge-sensitive D flip-flops.

3. **Reset strategy**

Let us use active-low asynchronous reset input reset_n. For reset_n = 0, the counter holds the previous output. For the reset_n = 1, the counter output increments on the rising edge of the clock.

4. **Let us document the entries in the state table**

The state table of the MOD-4 synchronous binary up-counter is shown in Table 7.5.

5. **Let us document the entries in the excitation table**

The excitation table consists of the information about the present state, next state and excitation input (Table 7.6).

Table 7.5 State table of the MOD-4 binary up-counter

Present state (q1 q0)	Next state ($q1^+$ $q0^+$)
00	01
01	10
10	11
11	00

Table 7.6 Excitation table of the MOD-4 counter

Present state (q1 q0)	Next state ($q1^+$ $q0^+$)	Excitation input (D1 D0)
00	01	01
01	10	10
10	11	11
11	00	00

6. **Let us deduce the logic at the input of the D1, D0**

Let us use the present state q1, q0 as the inputs to get the logic at the D1, D0. Let us use the 2-variable K-map (Figs. 7.7 and 7.8).

$$D1 = \overline{q1} \cdot q0 + \overline{q0} \cdot q1$$

$$D1 = q1 \oplus q0$$

$$D0 = \overline{q0}$$

Fig. 7.7 K-map to get the logic at D1

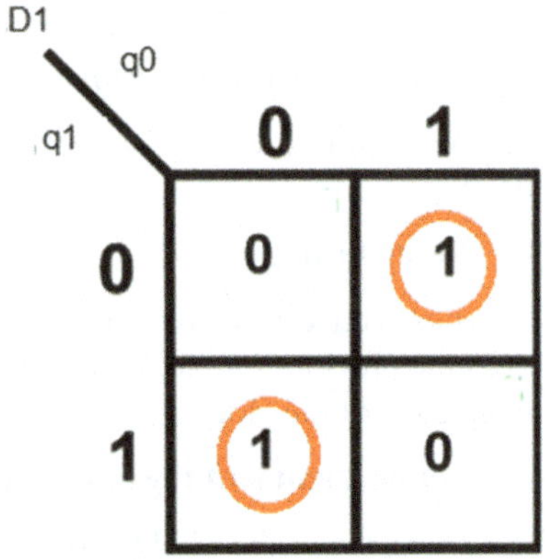

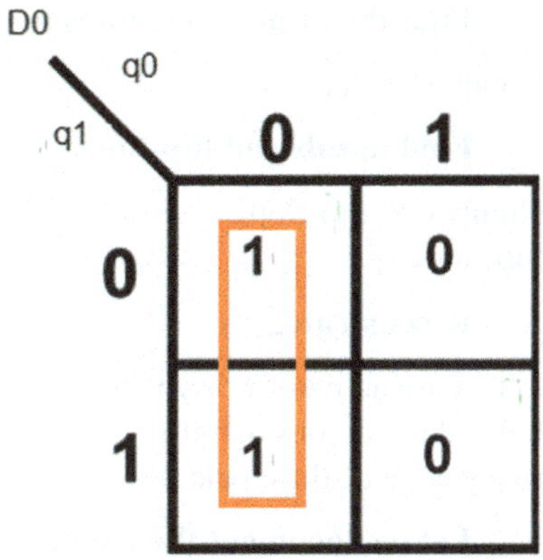

Fig. 7.8 K-map to get the logic at D0

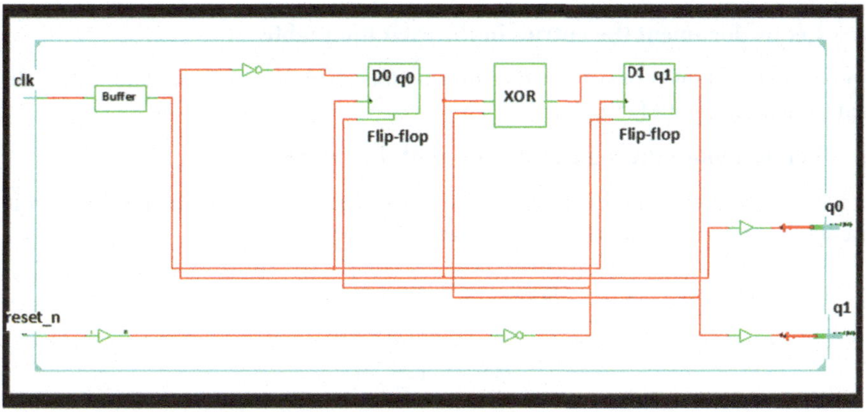

Fig. 7.9 2-bit synchronous binary up-counter

7. **Let us sketch the logic**

As discussed in the previous steps, we need to have two flip-flops and XOR gate to implement the MOD-4 synchronous binary up-counter (Fig. 7.9).

7.7 Exercise 1: Design of the Synchronous Down-Counters

Design the synchronous binary down-counter having active-low asynchronous reset (reset_n), rising edge-sensitive D flip-flops and the minimum logic gates.

Solution: Now, let us try to use the design foundation discussed in the previous sections to design the synchronous modulo-4 or MOD-4 binary down-counter. As name itself indicates, the counter has the four states. The four states of counter are so, s1, s2, s3 and for the binary down-counter design use the following steps.

1. **Find the number of states to get the counter**

Number of states = 4.

2. **Find number of flip-flops**

Number of flip-flops $= n = \log_2 4 = 2$. We will use the positive edge-sensitive D flip-flops.

3. **Reset strategy**

Let us use active-low asynchronous reset input reset_n. For reset_n = 0, the counter holds the previous output. For the reset_n = 1, the counter output decrements on the rising edge of the clock.

4. **Let us document the entries in the state table**

The state table of the MOD-4 synchronous binary down-counter is shown in Table 7.7.

5. **Let us document the entries in the excitation table**

The excitation table consists of the information about the present state, next state and excitation input (Table 7.8).

6. **Let us deduce the logic at the input of the D1, D0**

Let us use the present state q1, q0 as the inputs to get the logic at the D1, D0. Let us use the 2-variable K-map (Figs. 7.10 and 7.11).

$$D1 = \overline{q1} \cdot \overline{q0} + q1 \cdot q0$$

$$D1 = \overline{q1 \oplus q0}$$

$$D0 = \overline{q0}$$

Table 7.7 State table of the MOD-4 binary down-counter

Present state (q1 q0)	Next state ($q1^+$ $q0^+$)
11	10
10	01
01	00
00	11

Table 7.8 Excitation table of the MOD-4 down-counter

Present state (q1 q0)	Next state ($q1^+$ $q0^+$)	Excitation input (D1 D0)
11	10	10
10	01	01
01	00	00
00	11	11

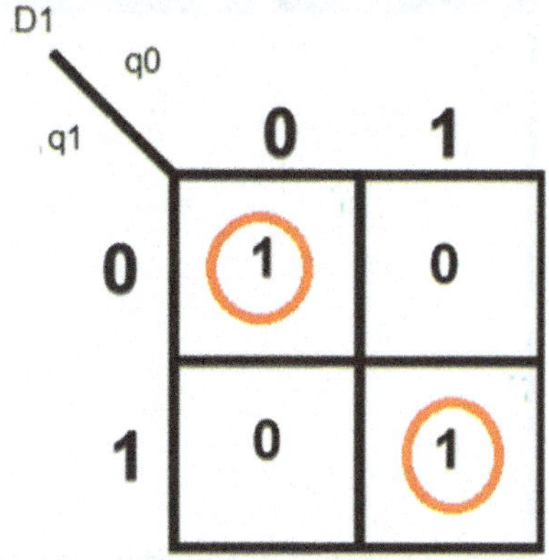

Fig. 7.10 K-map to get the logic at D1

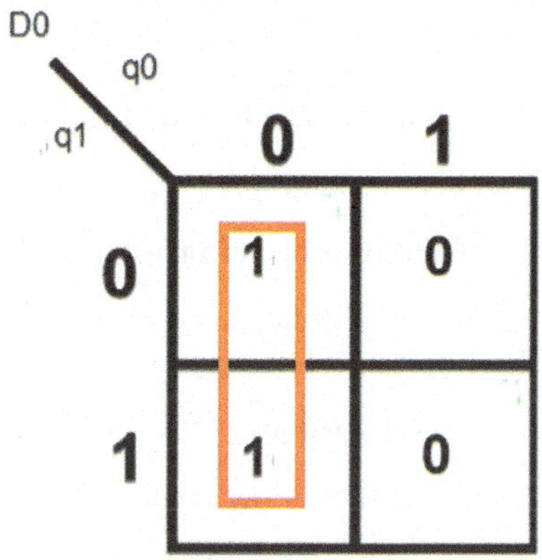

Fig. 7.11 K-map to get the logic at D0

7. **Let us sketch the logic**

As discussed in the previous steps, we need to have two flip-flops and XNOR gate to implement the MOD-4 synchronous binary down-counter (Fig. 7.12).

7.8 Exercise 2: Design of the Synchronous Gray Counter

Design the synchronous gray counter having active-low asynchronous reset (reset_n), rising edge-sensitive D flip-flops and the minimum logic gates.

Solution: Now, let us try to use the design foundation discussed in the previous sections to design the synchronous gray counter. As name itself indicates, the counter has the four states. so, s1, s2, s3 and for the gray counter design use the following steps.

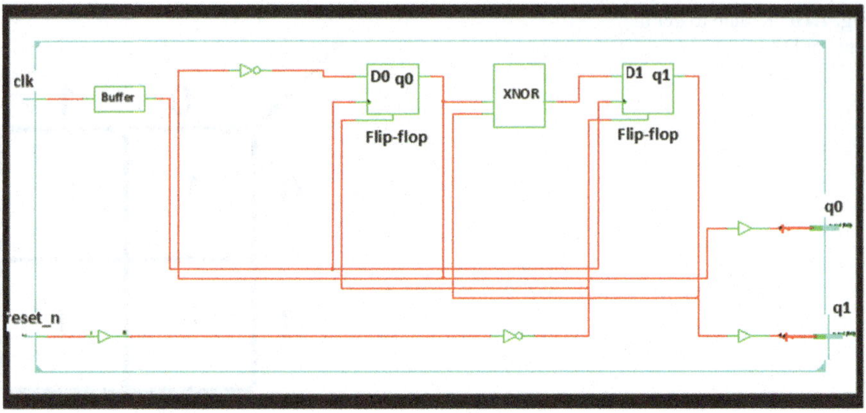

Fig. 7.12 2-bit synchronous binary down-counter

1. **Find the number of states to get the counter**

Number of states = 4.

2. **Find number of flip-flops**

Number of flip-flops = $n = \log_2 4 = 2$. We will use the positive edge-sensitive D flip-flops.

3. **Reset strategy**

Let us use active-low asynchronous reset input reset_n. For reset_n = 0, the counter holds the previous output. For the reset_n = 1, the counter output decrements on the rising edge of the clock.

4. **Let us document the entries in the state table**

The state table of the synchronous gray counter is shown in Table 7.9.

5. **Let us document the entries in the excitation table**

The excitation table consists of the information about the present state, next state and excitation input (Table 7.10).

Table 7.9 State table of the 2-bit gray counter

Present state (q1 q0)	Next state ($q1^+$ $q0^+$)
00	01
01	11
11	10
10	00

Table 7.10 Excitation table of the 2-bit gray counter

Present state (q1 q0)	Next state (q1$^+$ q0$^+$)	Excitation input (D1 D0)
00	01	01
01	11	11
11	10	10
10	00	00

6. **Let us deduce the logic at the input of the D1, D0**

Let us use the present state q1, q0 as the inputs to get the logic at the D1, D0. Let us use the 2-variable K-map (Figs. 7.13 and 7.14).

$$D1 = q0$$

$$D0 = \overline{q1}$$

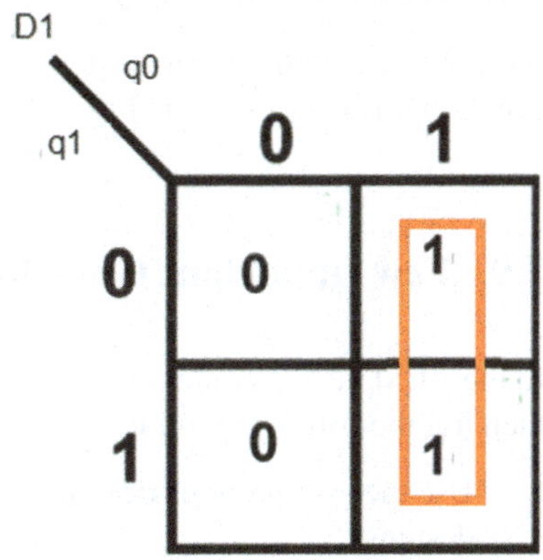

Fig. 7.13 K-map to get the logic at D1

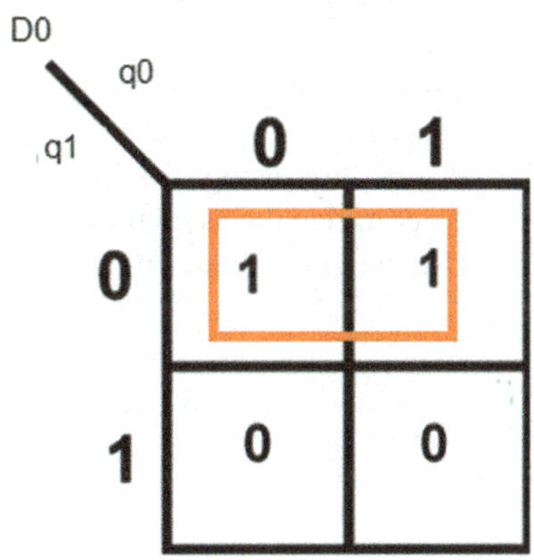

Fig. 7.14 K-map to get the logic at D0

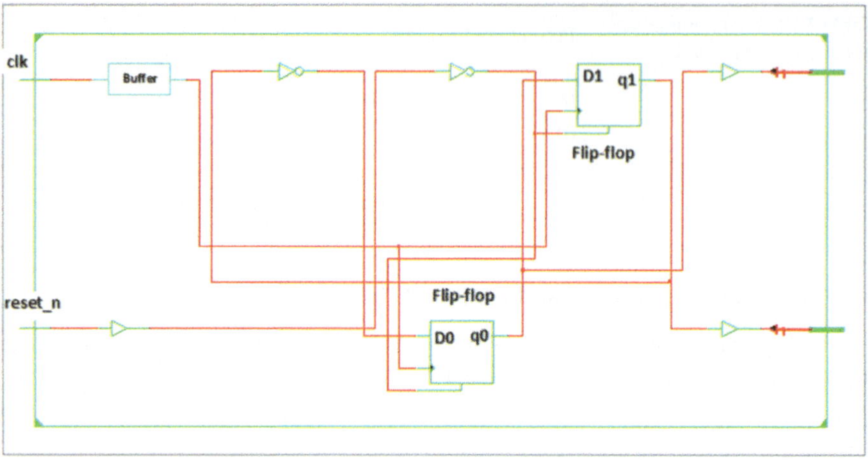

Fig. 7.15 2-bit gray counter

7. **Let us sketch the logic**

As discussed in the previous steps, we need to have two flip-flops to implement the synchronous gray counter (Fig. 7.15).

7.9 Few Important Guidelines

In the VLSI design context if you are FPGA engineer or the system design engineer, then use the following guidelines.

1. Use the synchronous design as they are faster as compared to the asynchronous designs.
2. Do not use the derived clocks as they have the clock skews.
3. Asynchronous clocking is prone to glitches so avoid use of the asynchronous counters and even asynchronous clocking.
4. Use the single PLL in the system for the single clock domain designs to derive the clock.
5. If you are using the asynchronous reset, then use the synchronizers to synchronize the asynchronous reset with the master reset.
6. Use the gray counter if the area and power requirement are minimum.

7.10 Important Takeaways

Following are few of the important points to conclude this chapter.

1. The synchronous counter uses the common clock for all the flip-flops.
2. In the asynchronous counters, the LSB flip-flop uses the main clock, and the clock of the MSB flip-flops is derived from the previous flip-flops stage.
3. Synchronous circuits are faster as compared to the asynchronous circuits.
4. For the reg-to-reg path, the maximum frequency is calculated by using the $fmax = \frac{1}{tpff+tcombo+tsu}$.
5. MOD-4 counter indicates the four states and two flip-flops to implement the design.
6. In the 2-bit gray counters, the combinational logic is not required and hence used to minimize the power and area.

Chapter 8
Important Design Scenarios

The sequential design techniques are useful during the system design to improve the design performance.

As discussed in Chap. 7, we can use the sequential design elements with the goal to have the maximum speed and lesser area. The chapter discusses the important design scenarios and techniques useful to design the sequential logic. The chapter is useful to understand about the duty cycle and how to design the sequential circuits with the goal to have duty cycle control.

8.1 MOD-3 Counter

Now, let us try to use the design techniques discussed in the previous chapters to design the synchronous modulo-3 or MOD-3 binary up-counter. As the name itself indicates, the counter has the three states so, s1 and s2, and for the binary up-counter design, use the following steps.

1. **Find the number of states to get the counter**

The number of states = 3.

2. **Find the number of flip-flops**

The number of flip-flops $= n = \log_2 3$. So we will use two positive edge sensitive D flip-flops.

3. **Reset strategy**

Let us use active-low asynchronous reset input reset_n. For reset_n=0, the counter outputs are initialized to zero. For the reset_n=1, the counter output increments on the rising edge of the clock.

V. Taraate, *Digital Design Techniques and Exercises*,
https://doi.org/10.1007/978-981-16-5955-3_8

4. **Let us document the entries in the state table**

The state table of the MOD-3 synchronous binary up-counter is shown in Table 8.1.

5. **Let us document the entries in the excitation table**

The excitation table consists of the information about the present state, next state and excitation input (Table 8.2).

6. **Let us deduce the logic at the input of the D1 and D0**

Let us use the present state q1 and q0 as the inputs to get the logic at the D1 and D0. Let us use the 2-varibale K-map (Figs. 8.1 and 8.2).

$$D1 = q0$$

$$D0 = \overline{q0} \cdot \overline{q1}$$

$$D0 = \overline{q1 + q0}$$

7. **Let us sketch the logic**

As discussed in the previous steps, we need to have two flip-flops and NOR gate to implement the MOD-3 synchronous binary up-counter (Fig. 8.3).

The timing waveform of the MOD-3 synchronous binary up-counter is shown in Fig. 8.4. As shown, the counter has three states s0, s1 and s2 and the output of the counter is 00, 01 and 10. If we use q1 to generate an output, then an output is 0 for two clock cycles and logic 1 for one clock cycle. The design output from MSB is single clock for three clock cycles; hence, divide by 3 or MOD-3 counter.

Duty cycle: The duty cycle is ratio of the on-time (Ton) and the T = Ton + Toff, where T is the clock period at output and Toff is off-time.

Table 8.1 State table of the MOD-3 binary up-counter

Present state (q1 q0)	Next state ($q1^+$ $q0^+$)
00	01
01	10
10	00

Table 8.2 Excitation table of the MOD-3 counter

Present state (q1 q0)	Next state ($q1^+$ $q0^+$)	Excitation input (D1 D0)
00	01	01
01	10	10
10	00	00

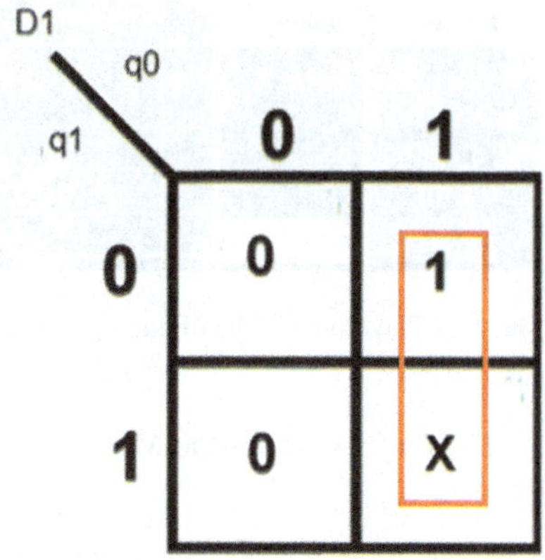

Fig. 8.1 K-map to get the logic at D1

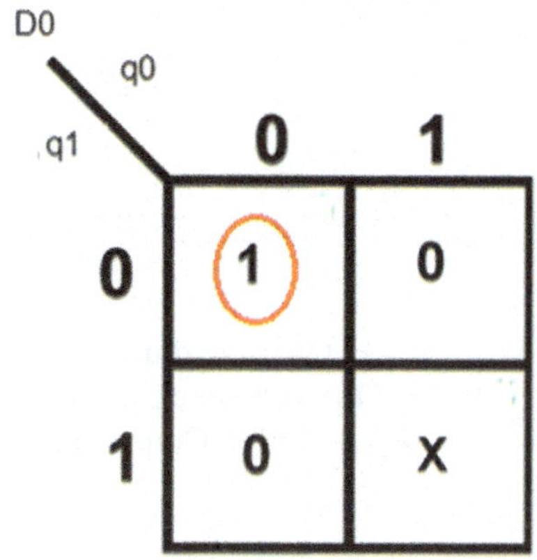

Fig. 8.2 K-map to get the logic at D0

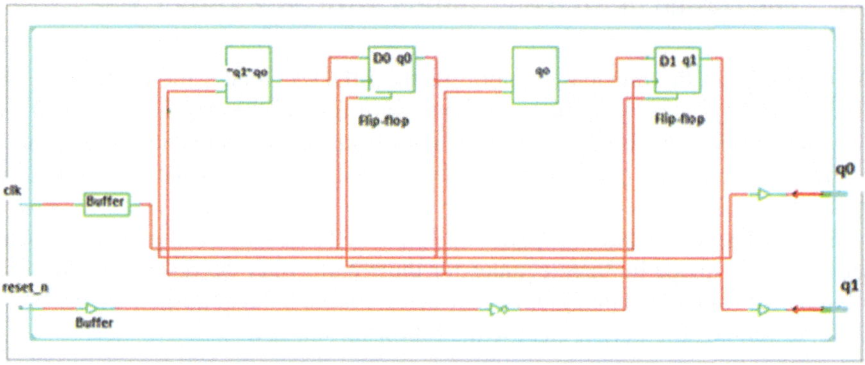

Fig. 8.3 Divide by three synchronous binary up-counter

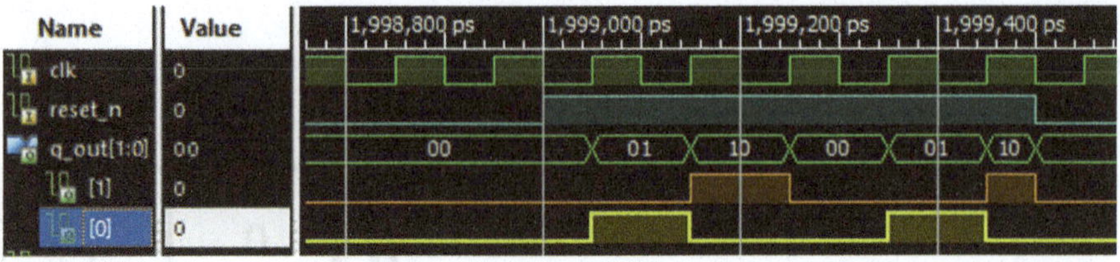

Fig. 8.4 Waveform of the divide by three synchronous binary up-counter

The duty cycle for the MSB output is given by

$$\begin{aligned}\text{Duty cycle} &= \frac{\text{Ton}}{\text{Ton} + \text{Toff}}\\ &= \frac{1}{1+2}\\ &= \frac{1}{3}\\ &= 0.3333\\ &= 33.33\%\end{aligned}$$

where Ton = On cycle time period
Toff = Off cycle time period
T = Ton + Toff = Output clock period.

8.2 The Design of MOD-3 Counter with 50% Duty Cycle

As discussed in the previous section, for the MOD-3 counter the output has the 33.33% duty cycle. Now, as the output has logic 1 for one clock cycle and logic 0 for two clock cycles, the use of the clock is not recommended in the design. For most of the synchronous designs, it is recommended to use the clock which has 50% duty cycle that is, Ton = Toff.

So, the design discussed in Sect. 8.1 should be tweaked by using the additional logic. For the divide by three counter for three half-cycle, the output should have logic 0, and for the remaining three half-clock cycles, the output should have logic 1. So, to get the 50% duty cycle, use the following strategy.

1. Design the MOD-3 counter as discussed in Sect. 8.1 using positive edge sensitive D flip-flops.
2. Get the output q1 which has 33.33% duty cycle.
3. Sample the q1 on the negative edge of the clock using negative edge sensitive D flip-flop.
4. Use OR gate to get the 50% duty cycle output q_out. The q_out = OR(q1, q1_n).

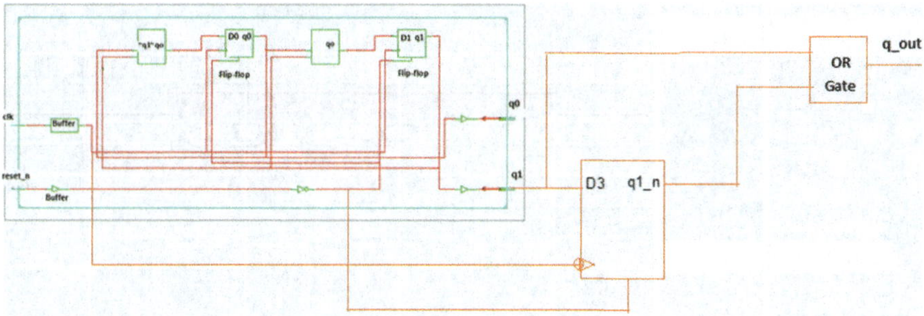

Fig. 8.5 Divide by three synchronous binary up-counter with 50% duty cycle

5. Sketch the schematic and observe the timing waveform.

The timing waveform of the MOD-3 synchronous binary up-counter is shown in Fig. 8.5. As shown, the counter has three states s0, s1 and s2 and the output of the counter is 00, 01 and 10. If we use q1 to generate an output, then an output is 0 for the one and half clock cycles and logic 1 for one and half clock cycle. The design output from MSB is single clock for three clock cycles hence divide by 3 or MOD-3 counter having 50% duty cycle.

Duty cycle: The duty cycle is ratio of the on-time (Ton) and the T = Ton + Toff, where T is the clock period at output and Toff is off-time.

The duty cycle for the MSB output is given by

$$\begin{aligned}\text{Duty cycle} &= \frac{\text{Ton}}{\text{Ton} + \text{Toff}}\\ &= \frac{1.5}{1.5 + 1.5}\\ &= \frac{1}{2}\\ &= 0.5000\\ &= 50.00\%\end{aligned}$$

where Ton = On cycle time period = Three half-cycles on

Toff = Off cycle time period = Three half-cycle off

T = Ton + Toff = Output clock period.

Here, duty cycle is 50% (Fig. 8.6).

8.3 Applications and Use of Counters

Most of the time, we use the synchronous counters in the system design to get the desired frequency output. The counters are frequency divider network and are used

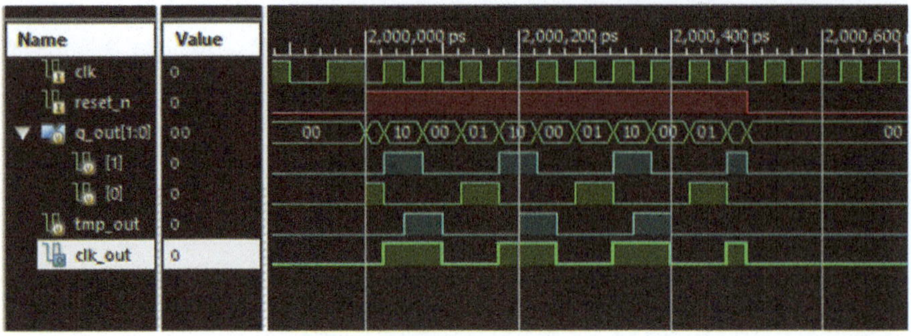

Fig. 8.6 Waveform of the divide by three synchronous binary up-counter with 50% duty cycle

to generate an output which is divided by n value. For example, MOD-16 counter has the 16 states s0 to s15 and is used to get the output as divided by 16. If input clock is 160 MHz, then the output of the MOD-16 counter is 10 MHz.

We can design the MOD-n synchronous up- or down-counters using the techniques discussed in Chap. 7. In the practical scenarios, we need to have few counters which can generate an output as 8, 4, 2, 1 or 0, 8, 12, 14, 15, 7, 3, 1, 0....

They are special counters and can be designed by observing the output pattern. This section discusses about the design of these counters.

8.3.1 Ring Counter

As name indicates, the ring counter generates an output sequence as 8, 4, 2, 1, 8, 4,..... . The counter can be designed very quickly using the state table. So, let us discuss the steps to design the ring counter.

1. **Find the maximum count**

For the counter to have output as 8, 4, 2, 1, the maximum count value is 8.

2. **Use the maximum count to find the number of flip-flops required**

In the binary, 8 is represented as 1000. Hence, the number of flip-flops required is equal to 4.

3. **Document the state table entries**

The counter has four states, and the entries are documented in Table 8.3.

4. **Document the excitation input of the flip-flops**

Let us document the excitation input to get the next state count value (Table 8.4).

5. **Observe the present state and next state to get the desired value as excitation to data input of flip-flops.**

Table 8.3 Four-bit ring counter state table

Present state (q3 q2 q1 q0)	Next state ($q3^+$ $q2^+$ $q1^+$ $q0^+$)
1000	0100
0100	0010
0010	0001
0001	1000

Table 8.4 Four-bit ring counter excitation table

Present state (q3 q2 q1 q0)	Next state ($q3^+$ $q2^+$ $q1^+$ $q0^+$)	Excitation input (D3 D2 D1 D0)
1000	0100	0100
0100	0010	0010
0010	0001	0001
0001	1000	1000

As shown in Table 8.5, the relationship between the present state and next state is on the rising edge of the clock, and the output of the ring counter shifts by 1-bit. So, using this let us get the Boolean equations to have D3, D2, D1, D0 inputs.

6. **Boolean equations**

Let us get the Boolean equations

$$D3 = Q0$$

Table 8.5 Excitation table to deduce the expressions

Present State (q3 q2 q1 q0)	Next State ($q3^+$ $q2^+$ $q1^+$ $q0^+$)	Excitation Input (D3 D2 D1 D0)
1000	0100	0100
0100	0010	0010
0010	0001	0001
0001	1000	1000

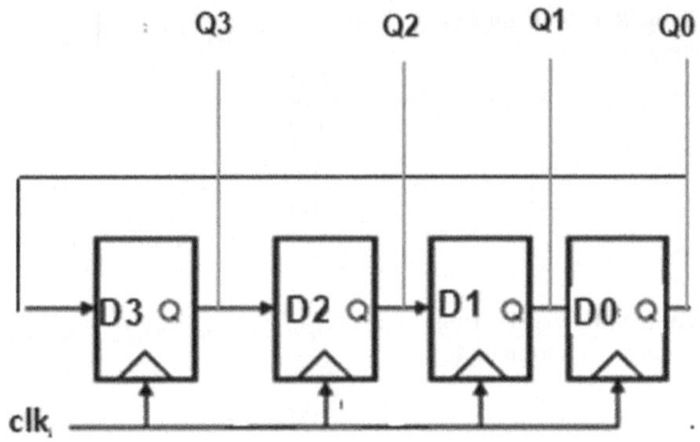

Fig. 8.7 Four-bit ring counter design

$$D2 = Q3$$

$$D1 = Q2$$

$$D0 = Q1$$

7. **Let us sketch the logic**

The design of the 4-bit ring counter is shown in Fig. 8.7.

8.3.2 Johnson Counter

Johnson counter is twisted ring counter and is used to generate an output sequence as 0, 8, 12, 14, 15, 7, 3, 1, 0, etc. The counter can be designed very quickly using the state table. So, let us discuss the steps to design the ring counter.

1. **Find the maximum count**

For the counter to have output as 0, 8, 12, 14, 15, 7, 3, 1, 0, etc., the maximum count value is 15.

2. **Use the maximum count to find the number of flip-flops required**

In the binary, 15 is represented as 1111. Hence, the number of flip-flops required is equal to 4.

3. **Document the state table entries**

The counter has four states, and the entries are documented in Table 8.6.

4. **Document the excitation input of the flip-flops**

Let us document the excitation input to get the next state count value (Table 8.7).

Table 8.6 Four-bit twisted ring counter state table

Present state (q3 q2 q1 q0)	Next state ($q3^{+}$ $q2^{+}$ $q1^{+}$ $q0^{+}$)
0000	1000
1000	1100
1100	1110
1110	1111
1111	0111
0111	0011
0011	0001
0001	0000

Table 8.7 Excitation table of twisted ring counter

Present state (q3 q2 q1 q0)	Next state ($q3^{+}$ $q2^{+}$ $q1^{+}$ $q0^{+}$)	Next state (D3 D2 D1 D0)
0000	1000	1000
1000	1100	1100
1100	1110	1110
1110	1111	1111
1111	0111	0111
0111	0011	0011
0011	0001	0001
0001	0000	0000

5. **Observe the present state and next state to get the desired value as excitation to data input of flip-flops.**

As shown in Table 8.8 the relationship between the present state and next state is on the rising edge of the clock, and the output of the twisted ring counter shifts by 1-bit, and the MSB of the counter is complement of the LSB output. So, using this, let us get the Boolean equations to have D3, D2, D1, D0 inputs.

6. **Boolean equations**

Let us get the Boolean equations

$$D3 = \overline{Q0}$$

$$D2 = Q3$$

$$D1 = Q2$$

$$D0 = Q1$$

Table 8.8 Excitation table to deduce the expressions

Present State (q3 q2 q1 q0)	Next State ($q3^+$ $q2^+$ $q1^+$ $q0^+$)	Next State (D3 D2 D1 D0)
0000	1000	1000
1000	1100	1100
1100	1110	1110
1110	1111	1111
1110	0111	0111
0111	0011	0011
0011	0001	0001
0001	0000	0000

7. **Let us sketch the logic**

The design uses the four flip-flops, and the counter is called as twisted ring counter (Fig. 8.8).

8.4 Exercises

Let us complete the exercises on the sequential design using the basic fundamentals and design techniques discussed.

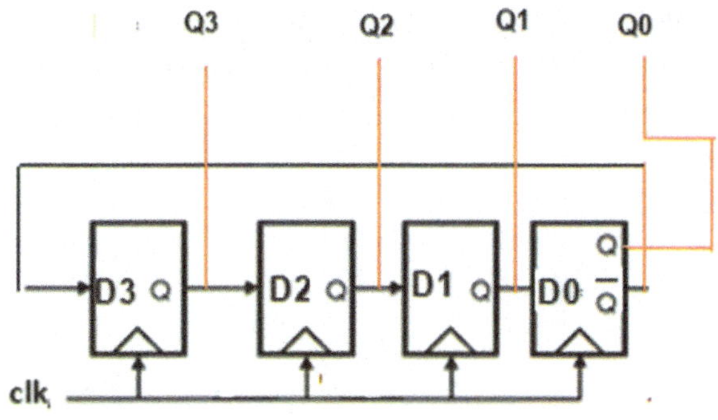

Fig. 8.8 Four-bit Johnson counter design

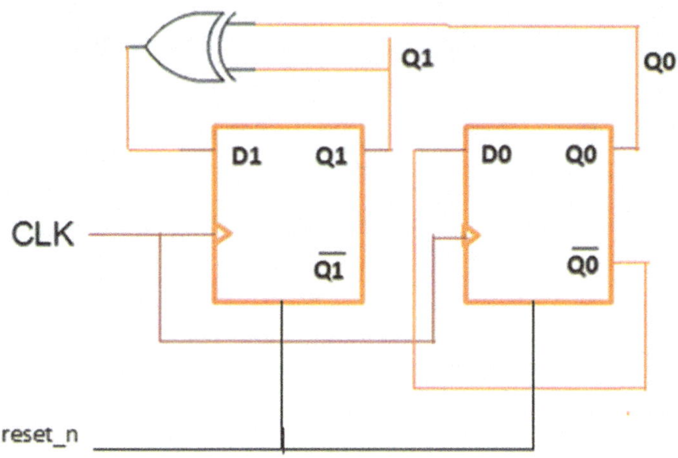

Fig. 8.9 Sequential design-1

8.4.1 *Exercise 1: The Counter Output*

For Fig. 8.9, find the output sequence.

Solution: During the reset_n = 0, the counter output Q1Q0 = 00. Consider the $Q1^+$, $Q0^+$ as next state.

So, let us document the next state in the table to get the output sequence.

As documented in Table 8.9, the sequence at Q1Q0 is 00, 01, 10, 11, 00, etc., and the design is synchronous MOD-4 binary up-counter.

8.4.2 *Exercise 2: Find the Output Sequence*

Find the output sequence for the design shown in Fig. 8.10.

Table 8.9 Sequence table for sequential design-1

CLK	Q1	Q0	$D1 = Q1 \oplus Q0$	$D0 = \overline{Q0}$	$Q1^+$	$Q0^+$
1	0	0	0	1	0	1
2	0	1	1	0	1	0
3	1	0	1	1	1	1
4	1	1	0	0	0	0

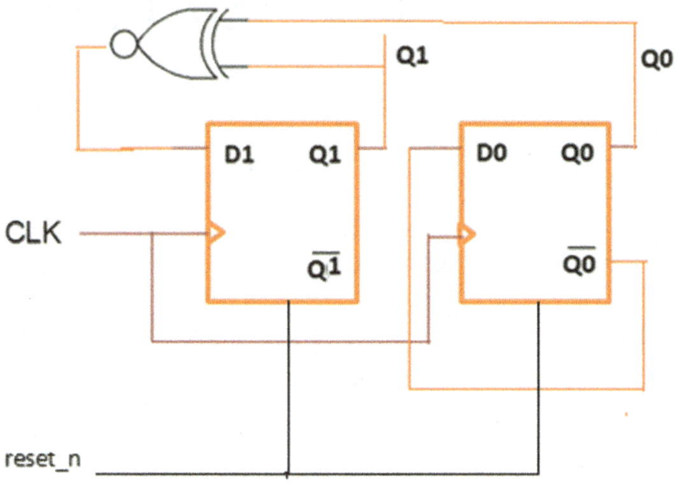

Fig. 8.10 Sequential design-2

Solution: During the reset_n = 0, the counter output Q1Q0 = 00. Consider the $Q1^+$, $Q0^+$ as next state.

So, let us document the next state in the table to get the output sequence.

As documented in Table 8.10, the sequence at Q1Q0 is 11, 10, 01, 00, 11, etc., and the design is synchronous MOD-4 binary down counter.

Table 8.10 Sequence table for sequential design-2

CLK	Q1	Q0	$D3 = \overline{Q1 \oplus Q0}$	$D0 = \overline{Q0}$	$Q1^+$	$Q0^+$
1	0	0	1	1	1	1
2	1	1	1	0	1	0
3	1	0	0	1	0	1
4	0	1	0	0	0	0

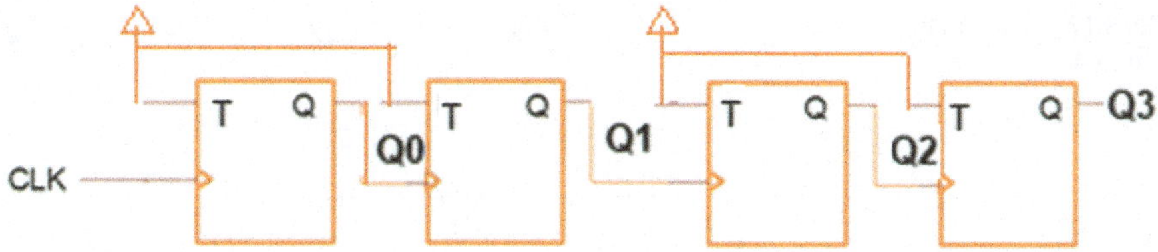

Fig. 8.11 Sequential design-3

8.4.3 Exercise 3: Operating Frequency of Design

Consider the CLK frequency as 100 MHz. What is frequency at outputs Q3 Q2 Q1 Q0?

Solution: As shown in Fig. 8.11, the design is asynchronous and LSB toggle flip-flop receives the clock (CLK). The toggle flip-flop toggles when $T = 1$ and generates an output as divide by 2 of CLK.

As the output of the LSB stage is used as the clock input to the MSB stage, each flip-flop output is divided by two sequential circuit.

The frequency at the output of each flip-flop is calculated as

1. $\text{Q0f} = \frac{\text{fCLK}}{2} = \frac{100\,\text{MHz}}{2} = 50\ \text{MHz}$
2. $\text{Q1f} = \frac{\text{fCLK}}{4} = \frac{100\,\text{MHz}}{4} = 25\ \text{MHz}$
3. $\text{Q2f} = \frac{\text{fCLK}}{8} = \frac{100\,\text{MHz}}{8} = 12.50\ \text{MHz}$
4. $\text{Q3f} = \frac{\text{fCLK}}{16} = \frac{100\,\text{MHz}}{16} = 6.25\ \text{MHz}.$

8.4.4 Exercise 4: Output on 1024th Clock Cycle

Consider the design shown in the figure. During the reset condition, the output Q3Q2Q1Q0 = 1000. What is output at 1024th clock?

Solution: As shown in Fig. 8.12, the design is synchronous 4-bit ring counter. The design uses the shifter, and the output Q0 is feedback to the D3. So, let us create the state table to get the output at 1024th clock pulse. Consider the present state as Q3Q2Q1Q0 and next state as $Q3^{+}Q2^{+}Q1^{+}Q0^{+}$.

As documented in Table 8.11, the design has output 1000 at 4th clock, so at 1024th clock, the design has output Q3Q2Q1Q0 = 1000.

8.4.5 Exercise 5: Output on the 4th Clock Cycle

Consider the design shown in Fig. 8.13. During the reset condition the output Q1Q0 = 00. What is output on the 4th clock?

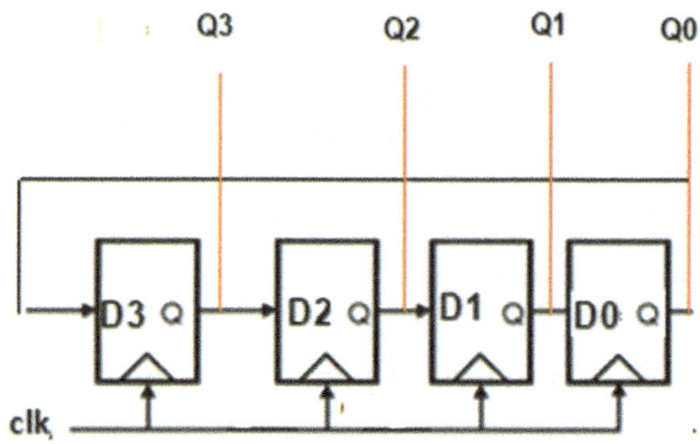

Fig. 8.12 Sequential design-4

Table 8.11 Truth-table of 4-bit ring counter

CLK	Q3	Q2	Q1	Q0	$Q3^{+}$	$Q2^{+}$	$Q1^{+}$	$Q0^{+}$
0	1	0	0	0	0	1	0	0
1	0	1	0	0	0	0	1	0
2	0	0	1	0	0	0	0	1
3	0	0	0	1	1	0	0	0

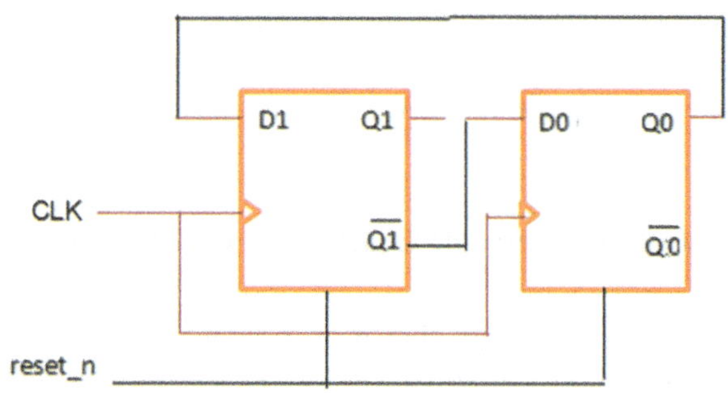

Fig. 8.13 Sequential design-5

Solution: As shown in Fig. 8.13, the design is synchronous. So, let us create the state table to get the output on 4th clock pulse. Consider the present state as Q1Q0 and next state as $Q1^{+}Q0^{+}$.

As documented in Table 8.12, the design is synchronous 2-bit gray counter and has output 00 on 4th clock.

8.4.6 Exercise 6: Output at 10th Clock Pulse

Consider the design shown in Fig. 8.14. During the reset condition, the output Q3Q2Q1Q0 = 0000. What is output at 10th clock?

Table 8.12 Sequence table for sequential design-5

CLK	Q1	Q0	$D1 = Q0$	$D0 = \overline{Q1}$	$Q1^+$	$Q0^+$
0	0	0	0	1	0	1
1	0	1	1	1	1	1
2	1	1	1	0	1	0
3	1	0	0	0	0	0

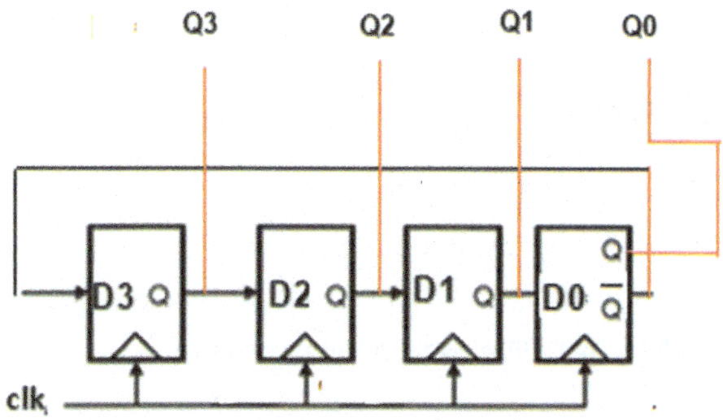

Fig. 8.14 Sequential design-6

Solution: As shown in Fig. 8.12, the design is synchronous 4-bit twisted ring counter. The design uses the shifter, and the output $\overline{Q0}$ is feedback to the D3. So, let us create the state table to get the output at 10th clock pulse. Consider the present state as Q3Q2Q1Q0 and next state as $Q3^+Q2^+Q1^+Q0^+$.

As documented in Table 8.13, the design has output 0000 at 8th clock so at 10th clock the design has output Q3Q2Q1Q0 = 1100.

Table 8.13 Truth-table of 4-bit twisted ring counter

CLK	Q3	Q2	Q1	Q0	$Q3^+$	$Q2^+$	$Q1^+$	$Q0^+$
0	0	0	0	0	1	0	0	0
1	1	0	0	0	1	1	0	0
2	1	1	0	0	1	1	1	0
3	1	1	1	0	1	1	1	1
4	1	1	1	1	0	1	1	1
5	0	1	1	1	0	0	1	1
6	0	0	1	1	0	0	0	1
7	0	0	0	1	0	0	0	0

Table 8.14 Truth-table of right shifter

CLK	Q3	Q2	Q1	Q0	$Q3^+$	$Q2^+$	$Q1^+$	$Q0^+$
1	0	0	0	0	d_in	0	0	0
2	d_in	0	0	0	0	d_in	0	0
3	0	d_in	0	0	0	0	d_in	0
4	0	0	d_in	0	0	0	0	d_in

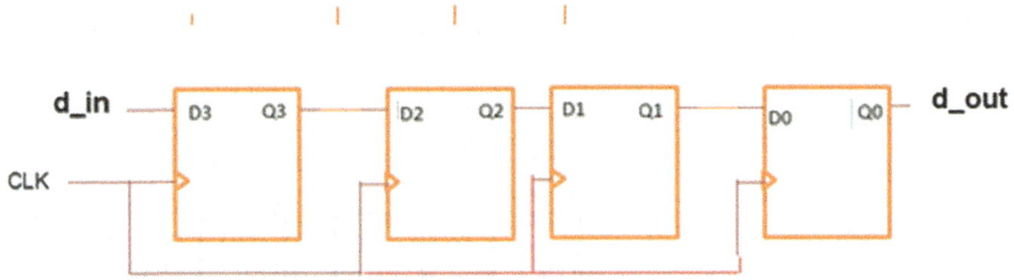

Fig. 8.15 Serial input serial output shift register

8.4.7 *Exercise 7: Design the Serial Input Serial Output Shift Register*

Using the D flip-flops, design the serial input serial output right shift operation. Consider the maximum latency as 4 clocks to get the output?

Solution: Let us create the state table to design the right shift operation. Consider the present state as Q3Q2Q1Q0 and next state as $Q3^+Q2^+Q1^+Q0^+$.

As documented in Table 8.14 to get the right shift operation, let us use the inputs of the flip-flops as D3 = d_in, D2 = Q3, D1 = Q2, D0 = Q1, and get the output d_out from Q0. The design is shown in Fig. 8.15.

8.5 Important Takeaways

The following are few of the important points to conclude this chapter.

1. The duty cycle is given by duty cycle $= \frac{\text{Ton}}{\text{Ton+Toff}}$.
2. It is recommended to have the design output with 50% duty cycle.
3. Ring counters and Johnson counter are special counters and used to repeat the sequence.
4. Johnson counter is also called as twisted ring counter.
5. The design should not have the mix of the positive edge and negative edge triggered flip-flops in the same path.
6. Avoid use of the asynchronous counters and have the glitch-free designs.
7. Using the shifters, we can design the ring and twisted ring counters.

Chapter 9
FSM Design Techniques

The understanding of the FSM designs and techniques is useful to develop the FSM-based designs and controllers.

The FSM is finite state machine and used to design the FSM controllers. For example, to detect the sequence of 1010 from the input, we can think of FSM designs. The arbitrary counters, sequence detectors and controllers can be designed using efficient FSM design techniques. In the previous chapters, we have discussed about the combinational and sequential design techniques. In this chapter, let us discuss about the FSM design techniques and their applications in the digital design.

9.1 What Is FSM?

FSM is finite state machine and used to design the controllers and sequence detectors. In most of the sequential designs, we need to design the control and timing unit. So, FSM design techniques are useful to design the efficient timing and control algorithms and designs.

The FSMs are classified mainly into two categories as follows.

1. Moore FSM
2. Mealy FSM

So now, let us discuss about the Moore and Mealy machines and their use in the design.

V. Taraate, *Digital Design Techniques and Exercises*,
https://doi.org/10.1007/978-981-16-5955-3_9

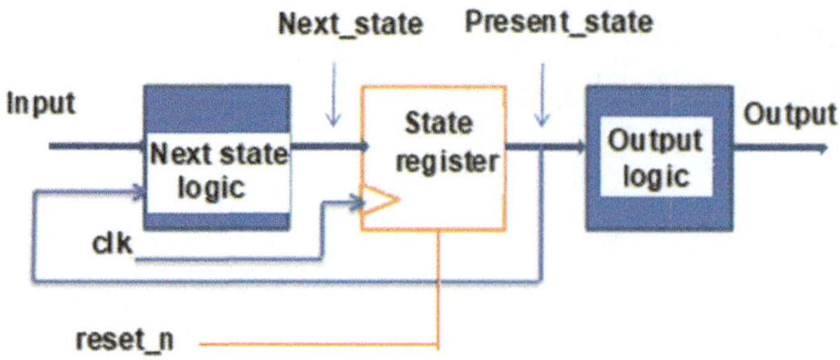

Fig. 9.1 Block diagram of Moore machine

9.1.1 Moore FSM

In the Moore FSM, an output is function of the present state only. As output is function of the present state, it is stable for one clock cycle. The state transition happens on the active edge of the clock.

The Moore FSM has three functional blocks as follows.

1. State register
2. Next state logic
3. Output logic.

The state register is sequential logic. The next state and output logic is combinational logic.

The block diagram of the Moore FSM is shown in Fig. 9.1.

As shown in Fig. 9.1, the Moore machine has three blocks, and our goal is to design the digital logic of these blocks.

9.1.2 Mealy FSM

In the Mealy FSM, an output is function of the present state and the inputs. As output is function of the present state and inputs, hence it may or may not be stable for one clock cycle. The state transition happens on the active edge of the clock. As compared to the Moore FSM, the Mealy FSM is prone to glitches.

The Mealy FSM has three functional blocks as follows.

1. State register
2. Next state logic
3. Output logic.

The state register is sequential logic. The next state and output logic is combinational logic.

The block diagram of the Mealy FSM is shown in Fig. 9.2.

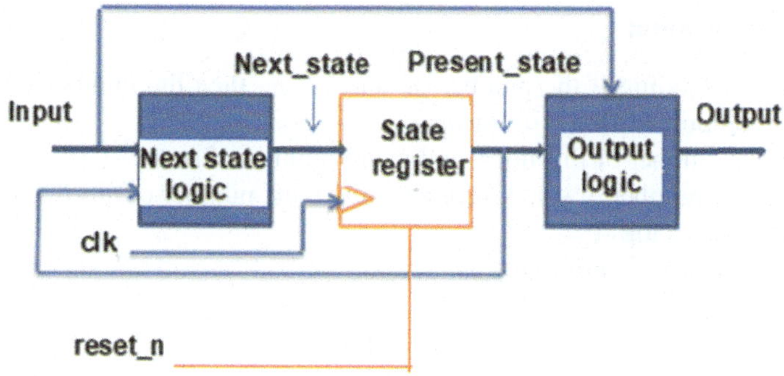

Fig. 9.2 Block diagram of Mealy machine

As shown in Fig. 9.2, the Mealy machine has three blocks, and our goal is to design the digital logic of these blocks.

9.1.3 Moore Versus Mealy FSM

The difference between the Moore and Mealy FSM is documented in Table 9.1.

9.2 State Encoding Methods

For the FSM designs, we should understand the state encoding. There are three types of the state encoding for FSM-based designs, and they are.

Table 9.1 Differences between Moore and Mealy machines

Moore machine	Mealy machine
Outputs are function of present state only	Outputs are function of the present state and inputs also
Output is stable for one clock cycle, and Moore FSM is not prone to glitches or spikes	Output may change multiple times depending on changes in the input and may or may not be stable for one clock cycle, and hence, Mealy FSM is prone to glitches or hazards
It requires more number of states as compared to Mealy machine	Mealy machine needs lesser states as compared to Moore machine
Moore FSM has the higher operating frequency as compared to Mealy machine	Mealy FSM has the lower operating frequency as compared to Moore machine

1. **Binary encoding**

In the binary encoding if the number of states is m, then the number of flip-flops required is computed using $n = \log_2 m$.

Where $n =$ number of flip-flops, and m are number of states.

Consider the number of states as $m = 4$, and then, number of flip-flops needed to design FSM is $n = \log_2 4 = 2$.

The four states are represented as

$$S0 = 00$$
$$S1 = 01$$
$$S2 = 10$$
$$S3 = 11$$

2. **Gray encoding**

In the gray encoding if the number of states are m, then the number of flip-flops required is computed using $n = \log_2 m$.

Where $n =$ number of flip-flops, and m are number of states.

Consider the number of states as $m = 4$, and then, number of flip-flops needed to design FSM is $n = \log_2 4 = 2$.

The four states are represented as

$$S0 = 00$$
$$S1 = 01$$
$$S2 = 11$$
$$S3 = 10$$

The gray encodingis useful in the FSM design to save the poweras in two successive gray numbers only one-bit changes.

3. **One-hot encoding**

In the one-hot encoding if the number of states is m, then the number of flip-flops required is computed using $n = m$. In this encoding, only one bit is active high or hot at a time.

Where $n =$ number of flip-flops, and m are number of states.

Consider the number of states as $m = 4$, and then, number of flip-flops needed to design FSM is $n = m = 4$.

The four states using one-hot encoding are represented as

$$S0 = 0001$$
$$S1 = 0010$$
$$S2 = 0100$$
$$S3 = 1000$$

The one-hot encoding is useful in the FSM design to have better timing if area is not the constraint.

9.3 Moore FSM Design

Now, let us design the FSM for the given specifications. What we need to do is that we need to design the digital logic for the

1. State register
2. Next state logic
3. Output logic

Let us consider the design of the sequential circuit to get the output data_out which is input clock frequency divided by 2 when data_in = 1. We can use the following design steps to design the Moore FSM.

1. **Find the number of states to get the divide by 2 output**

Number of states = 2. The states are $s0$ and $s1$.

2. **State diagram**

The state transition happens when data_in = 1 only (Fig. 9.3).

3. **Find number of flip-flops**

We will use the binary encoding and the number of flip-flops = $n = \log_2 2 = 1$. We will use the positive edge-sensitive D flip-flops.

4. **Reset strategy**

Let us use active-low asynchronous reset input reset_n. For reset_n = 0, the counter output is logic 0. For the reset_n = 1, the output increments or toggles on the rising edge of the clock.

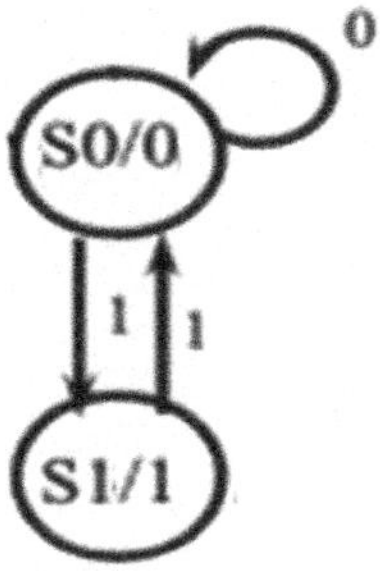

Fig. 9.3 Moore FSM of toggle flip-flop

Table 9.2 State table of the MOD-2 binary up-counter

Enable (data_in)	Present state ($q0$)	Next state ($q0^{+}$)
1	$s0$	$s1$
1	$s1$	$s0$
0	$s0$	$s0$
0	$s1$	$s1$

Table 9.3 Excitation table of the MOD-2 counter

Enable (data_in)	Present state ($q0$)	Next state ($q0^{+}$)	Excitation input ($D0$)
1	$s0 = 0$	$s1 = 1$	1
1	$s1 = 1$	$s0 = 0$	0
0	$s0 = 0$	$s0 = 0$	0
0	$s1 = 1$	$s1 = 1$	0

5. **Let us document the entries in the state table to get state register logic**

The state table of the MOD-2 synchronous counter is shown in Table 9.2.

So, we need to have single D flip-flop having asynchronous active-low reset input reset_n and active-high enable data_in.

6. **Use of the excitation table to design the next state logic**

The excitation table consists of the information about the present state, next state and excitation input (Table 9.3).

Now, let us use the *K*-map to deduce the Boolean equation for the next state (Fig. 9.4).

$$q0^{+} = \text{data\_in}.\overline{q0}$$

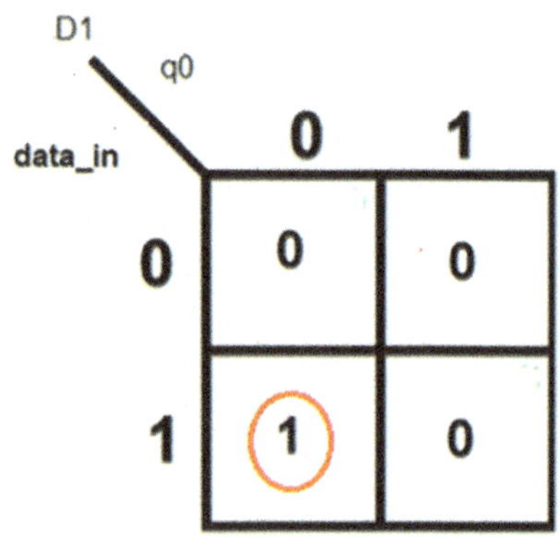

Fig. 9.4 K-map for the next state logic

Table 9.4 Truth-table of the output logic

Enable (data_in)	Present state ($q0$)	Output (data_out)
1	$s0 = 0$	$s0 = 0$
1	$s1 = 1$	$s1 = 1$
0	$s0 = 0$	$s0 = 0$
0	$s1 = 1$	$s1 = 1$

7. **Output logic**

In the Moore FSM output is function of the present state only. For this design, there is no any need of the output logicTable 9.4.

8. **Let us sketch the FSM design**

As discussed in the previous steps, we need to have flip-flop and logic gates to implement the MOD-2 counter which is divide by 2. Only care should be taken that the flip-flop should have active-high enable. As shown in Fig. 9.5, the state register uses data_in = 1 as active-high enable. During the data_in = 0, the output of the register is same as the present state.

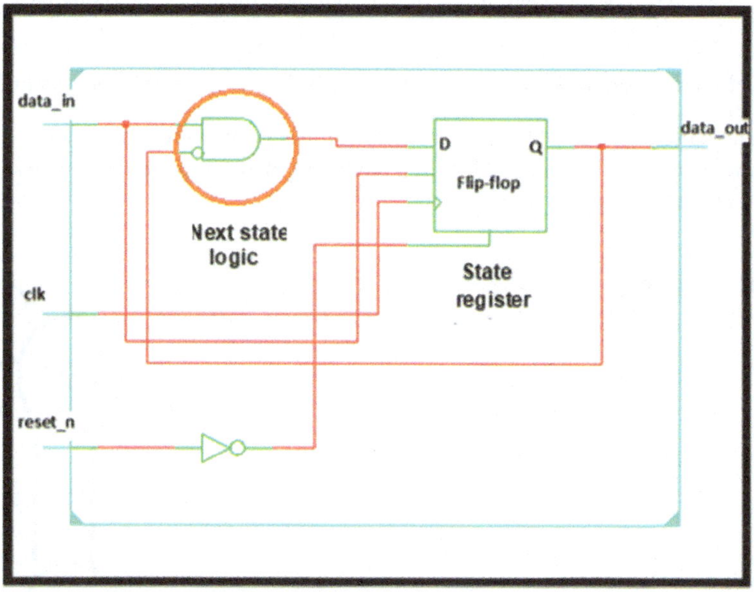

Fig. 9.5 MOD-2 counter having active-high enable

9.4 Mealy FSM Design

Let us consider the design of the sequential circuit to get the output data_out which is input clock frequency divided by 2 when data_in = 1. We can use the following design steps to design the Mealy FSM.

1. **Find the number of states to get the divide by 2 output**

Number of states = 2. The states are $s0$ and $s1$.

2. **State diagram**

The state transition happens when data_in = 1 only (Fig. 9.6).

3. **Find number of flip-flops**

We will use the binary encoding and the number of flip-flops = n= $\log_2 2 = 1$. We will use the positive edge-sensitive D flip-flops.

4. **Reset strategy**

Let us use active-low asynchronous reset input reset_n. For reset_n = 0, the counter output is logic 0. For the reset_n = 1, the output increments on the rising edge of the clock.

5. **Let us document the entries in the state table to get state register logic**

The state table of the MOD-2 synchronous counter is shown in Table 9.5.

So, we need to have single D flip-flop having asynchronous active-low reset input reset_n and active-high enable data_in.

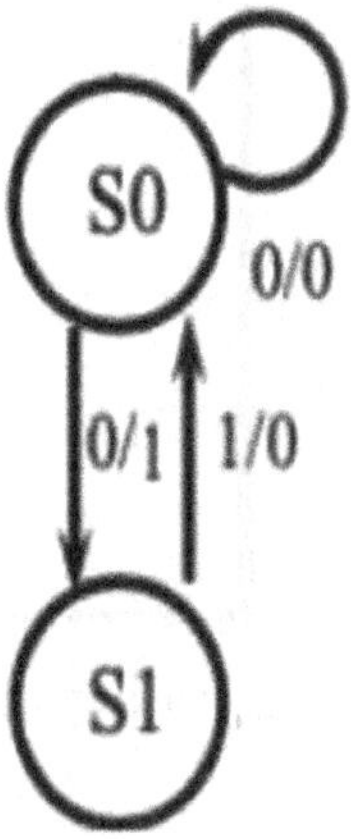

Fig. 9.6 Mealy Design of toggle flip-flop

Table 9.5 State table of the MOD-2 binary up-counter

Enable (data_in)	Present state ($q0$)	Next state ($q0^{+}$)
1	$s0$	$s1$
1	$s1$	$s0$
0	$s0$	$s0$
0	$s1$	$s1$

6. **Use of the excitation table to design the next state logic**

The excitation table consists of the information about the present state, next state and excitation input (Table 9.6).

Now, let us use the K-map to deduce the Boolean equation for the next state logic (Fig. 9.7).

$$q0^{+} = \text{data\_in}.\overline{q0}$$

7. **Output logic**

In the Mealy FSM output is function of the present state and changes in inputs. (Table 9.7).

Now, let us use the K-map to deduce the Boolean equation for the output logic (Fig. 9.8).

$$\text{data\_out} = \text{data\_in}.\overline{q0}$$

So, we need to have AND gate as an output logic.

Table 9.6 Excitation table of the MOD-2 counter

Enable (data_in)	Present state ($q0$)	Next state ($q0^{+}$)	Excitation input ($D0$)
1	$s0 = 0$	$s1 = 1$	1
1	$s1 = 1$	$s0 = 0$	0
0	$s0 = 0$	$s0 = 0$	0
0	$s1 = 1$	$s1 = 1$	0

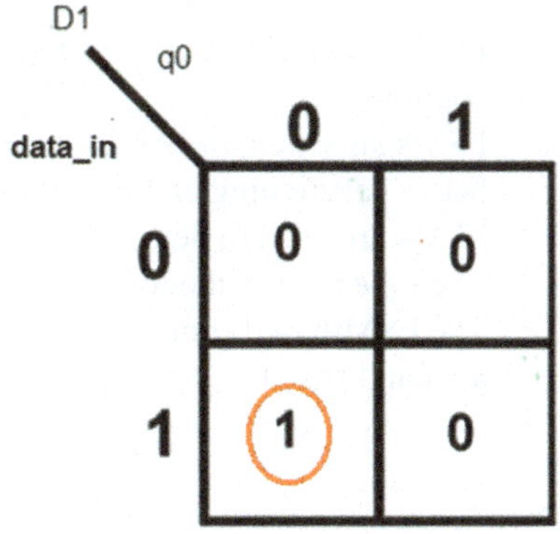

Fig. 9.7 K-map for the next state logic

Table 9.7 Truth-table of the output logic

Enable (data_in)	Present state ($q0$)	Output (data_out)
1	$s0 = 0$	0
1	$s1 = 1$	1
0	$s0 = 0$	0
0	$s1 = 1$	0

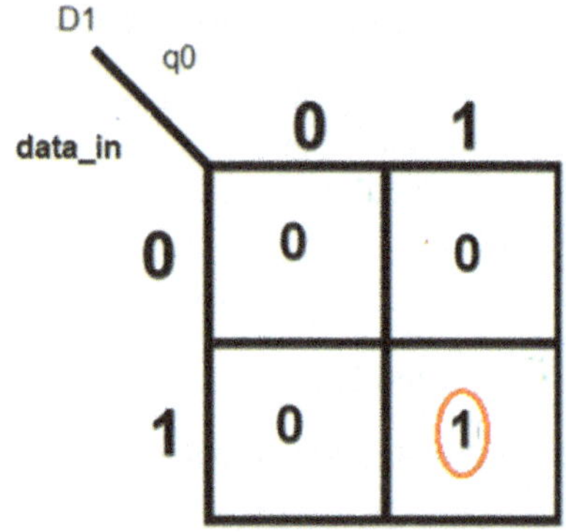

Fig. 9.8 K-map for the output logic

8. **Let us sketch the FSM Design**

As discussed in the previous steps, we need to have flip-flop and logic gates to implement the MOD-2 counter which is divide by 2. Only care should be taken that the flip-flop should have active-high enable. As shown in Fig. 9.9, the state register uses data_in = 1 as active-high enable. During the data_in = 0, the output of the register is same as the present state (Fig. 9.9).

9.5 Applications and Design Strategies

The FSM designs are useful to design the sequential logic with the objective to have better data and control path optimization. Following are few of the applications and the strategies while designing the FSM-based designs.

1. FSM-based design approach is used to design the arbitrary counters or random counter.
2. FSMs are extensively useful to design the large density counting circuit to have better partitioning and better area, timing.
3. FSMs are used to detect the sequence from the input stream.
4. The objective of the logic designer is to design the glitch-free FSMs.
5. The FSM-based controllers should have separate data and control path for better area and speed.

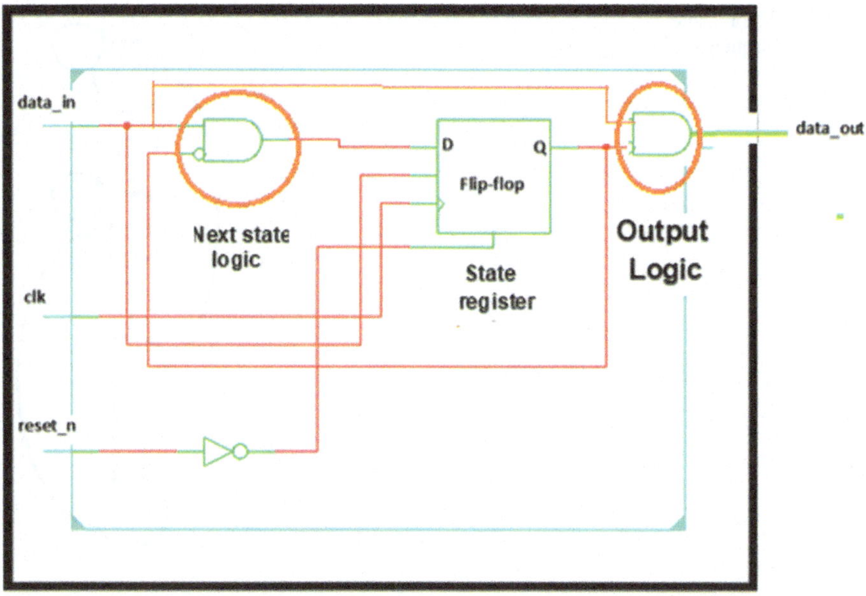

Fig. 9.9 MOD-2 counter having active-high enable

6. While designing the FSM to optimize for the area, try to have the strategy to eliminate unwanted or unused states.
7. Use the gray encoding for the power optimization.
8. If the area is not a constraints, then for the better timing, use the one-hot encoding.

9.6 Exercises

By using the techniques discussed in the previous sections, let us complete exercises.

9.6.1 Exercise 1: Moore Machine State Diagram

Sketch the Moore state diagram to detect the 101-overlapping sequence.

Solution: To detect the sequence 101, use the understanding of the Moore machine. Output is function of the present state only, and output is 1 when the sequence 101 is detected.

Consider the default state is *S*0 and output is 0. So, the state transition should be as follows.

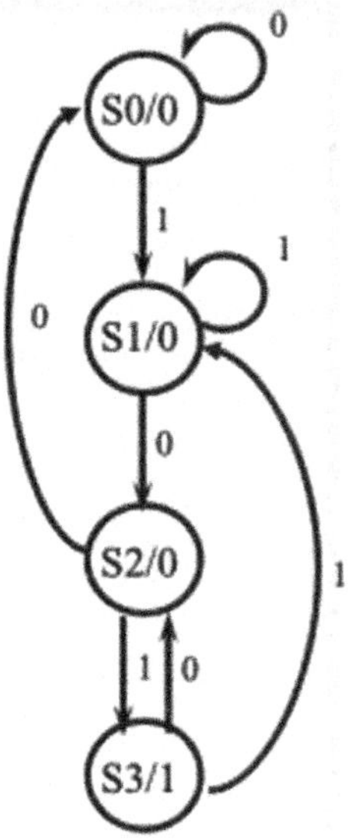

Fig. 9.10 Moore state diagram of sequence 101

1. If input is 1, then *S*0–*S*1. Be there in state *S*0 for input is equal to 0.
2. If next input bit is 0, then *S*1–S2. Be there in state *S*1 when input is 1.
3. If next input bit is 1, then *S*2–*S*3, and output is 1. If input is 0, then *S*2–*S*0.
4. To detect the overlapping sequence 101, if again next input is 0, then *S*3–*S*2. If input is 1, then state transition from *S*3–*S*1.

The state diagram is shown in Fig. 9.10.

9.6.2 Exercise 2: Mealy Machine

Sketch the Mealy state diagram to detect the 101-overlapping sequence.

Solution: To detect the sequence 101, use the understanding of the Mealy machine. Output is function of the present state and input also, and output is 1 when the sequence 101 is detected.

Consider the default state is *S*0 and output is 0. So, the state transition should be as follows.

1. If input is 1, then *S*0–*S*1. Be there in state *S*0 for input is equal to 0.
2. If next input bit is 0, then *S*1–*S*2. Be there in state *S*1 when input is 1.
3. To detect the overlapping sequence 101, if next input bit is 1, then *S*2–*S*1, and output is 1. If input is 0, then *S*2–*S*0.

The state diagram is shown in Fig. 9.11.

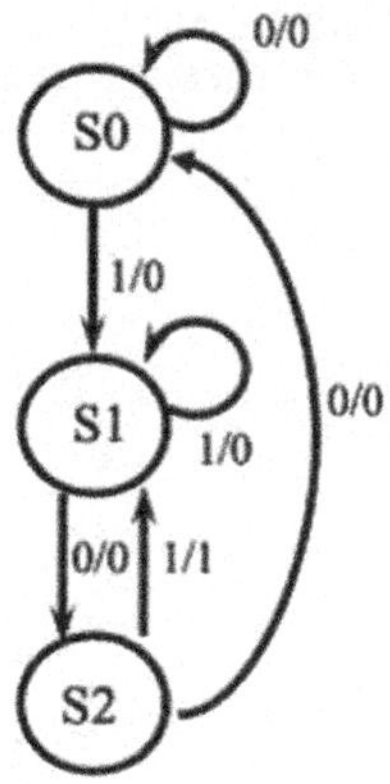

Fig. 9.11 **Mealy** state diagram of sequence 101

9.6.3 Exercise 3: One-Hot Encoding

For the following state diagram to detect the 101-overlapping sequence, document the states using one-hot encoding. How many flip-flops are needed to implement the one-hot encoding state machine? (Fig. 9.12).

Solution: For the state machine design using one-hot encoding, the number of flip-flops is equal to number of states. For the given Moore sequence detector, the number of flip-flops = 4.

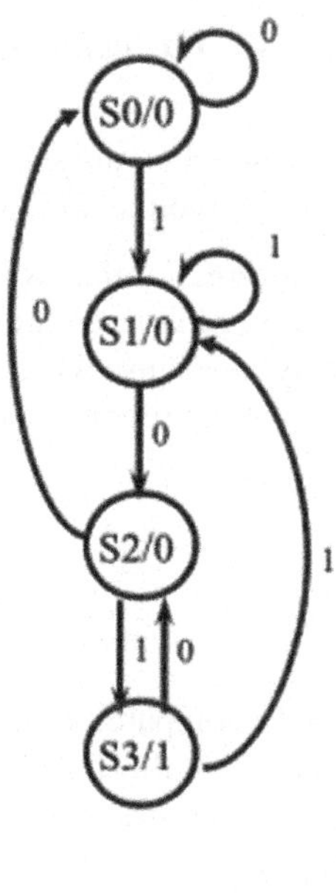

Fig. 9.12 Moore state diagram of sequence 101

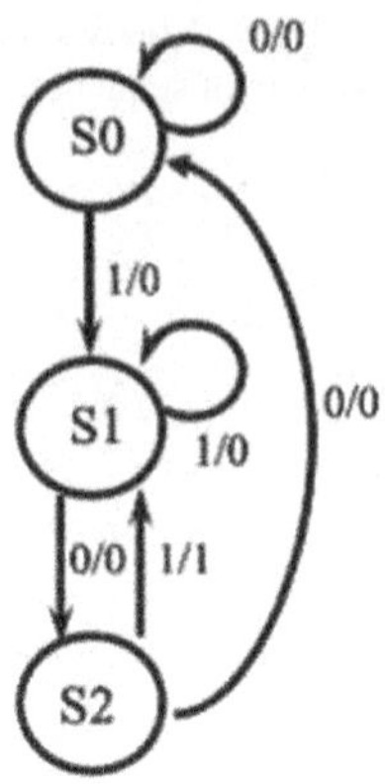

Fig. 9.13 Mealy state diagram of sequence 101

We can represent the states using one-hot encoding as

$$S0 = 0001$$
$$S1 = 0010$$
$$S2 = 0100$$

$$S3 = 1000$$

9.6.4 Exercise 4: FSM Area and Power Optimization

For the following state diagram to detect the 101-overlapping sequence, how we can think about the area and power optimization? (Fig. 9.13).

Solution: For the given Mealy sequence detector, the number of flip-flops = 2 if we use the binary or gray encoding. As there is single bit change in the two consecutive gray codes, use the gray encoding to optimize for the power.

We can represent the states using gray encoding as

$$S0 = 00$$
$$S1 = 01$$
$$S2 = 11$$

As compared to the one-hot encoding, the gray encoding optimizes the area as number of flip-flops for gray encoding are $\log_2 States$.

If states are four, then the number of flip-flops is 2. For three states, we need two flip-flops.

9.7 Important Takeaways

Following are few of the important points to conclude this chapter.

1. In the Moore FSM, an output is function of the present state only.
2. In the Mealy FSM, an output is function of the present state and inputs.
3. Mealy FSM needs lesser number of states as compared to Moore FSM.
4. The FSM-based design approach is useful to design the sequence detectors and random counters.
5. The FSM can use one of the state encoding method, binary, gray or one-hot.
6. Using the gray encoding, the power can be optimized.
7. Using the one-hot encoding method, the timing can be improved, but it increases the area.

Chapter 10
Advanced Design Techniques-1

The advanced digital design techniques are useful to improve the speed of design and even to optimize for the area and power.

Already in previous nine chapters, we have discussed the various design techniques. The chapter discusses the data and control path designs and the timing of the synchronous sequential circuits. Even this chapter focuses on the various advanced design techniques which are useful to improve design speed and to optimize for the area, and power. You can use these techniques in the design of the architecture and also in the high-speed digital designs.

10.1 Various Paths in the Design

In most of the designs, we have the sequential and combinational elements. If we consider the processor logic then we have the

1. ALU
2. Internal memory
3. Serio IO control
4. Interrupt control
5. Register array
6. Bus interface units
7. Control and timing unit
8. Clock and reset logic.

If the processor has the single clock domain, then the single clock is used as clock input for the various units.

Practically, the design has the

1. Clock paths
2. Reset paths
3. Data paths and control paths

V. Taraate, *Digital Design Techniques and Exercises*,
https://doi.org/10.1007/978-981-16-5955-3_10

Few of the recommendations during the design are

- The design should have the better clock and reset management schemes. For example, if the asynchronous reset is used then there should be the reset synchronizers to synchronize the internal asynchronous reset with the master reset.
- The clock distribution schemes should be such that there is uniform skew across the clock path.
- Don't generate the clocks using multiple sources as there is issue due to multiple clock domain and data convergence.
- If multiple clocks is requirement for the design, then use the synchronizers in the data and control paths.
- Use the level synchronizers to pass the control signals from one of the clock domain to another clock domain.
- Use the FIFO synchronizers to pass the data between the clock domains.
- Have the better data and control path management.

For the complex design, the data path and control path are discussed in the following sections.

10.2 Data and Control Paths

It is recommended to have the better data and control path logic. The data path logic allows the use of the input data and processing of the data depending on the control inputs. The main intention of the designer is to isolate the larger density data paths or to include the parallelism in the design to process the data depending on the control inputs.

So, the better strategy to design the data path logic is

1. Try to understand the data processing requirements. For example, consider the DSP processor performs the multiplication operation on 16-bit data, and it generates the 32-bit result (Fig. 10.1).
2. The main strategy is to use the design which can perform the multiplication operation using the data path logic.
3. Here the data path logic uses the 16-bit inputs as op_1, op_2 to generate an output as 32-bit result.

The role of the control path logic

The control path logic is used to generate the control and timing information to the data path logic. Even the control unit polls the handshaking signal status. The handshaking signal is used to convey the information to the control logic that the multiplication operation is performed, and now for the next control signal the data path logic can perform another multiplication operation.

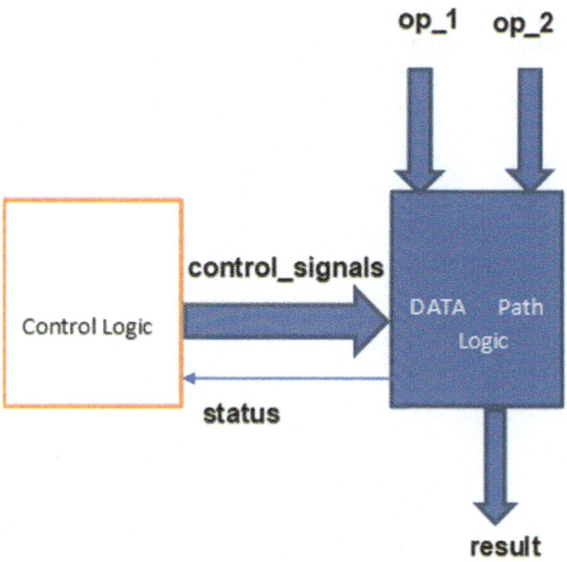

Fig. 10.1 Data and control path logic

Following are few important objectives to have the separate control path logic.

1. The control path logic can communicate the control and timing information to the data path logic.
2. Depending on the control signal information, the data path logic can perform the operation and generates the status signal as handshaking signal.
3. Even the control path logic is used to generate the control and timing information for the design to process the data.

Most of the time, we need to have better control and data paths for efficient area and timing of the design. Now let us discuss the sequence detector and its use as control logic in the control path.

10.3 Mealy Sequence Detector Design

Let us consider the design of the sequence detector to detect the sequence 101 from the input. If input is 1010100100101…then expected output is 0010100000001. We can use the design steps discussed in Chap. 9 to design the Mealy sequence detector

1. **State diagram**

The state transition happens when the input 1,0,1 at data_in (Fig. 10.2).

2. **Find the number of states to detect the sequence**

Number of states = 3. The states are $s0$ and $s1$ and $s2$.

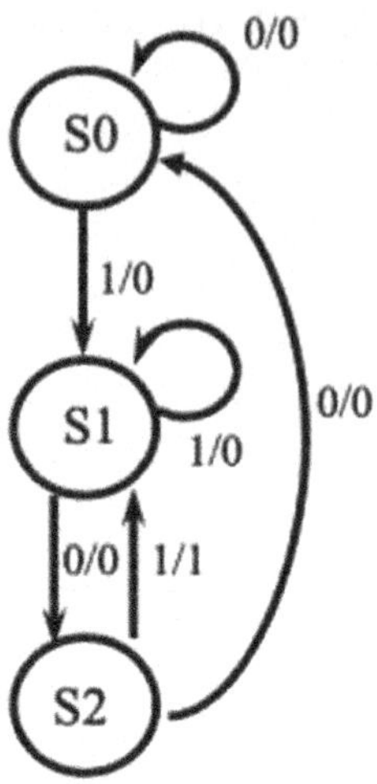

Fig. 10.2 Mealy FSM sequence detector

3. **Find number of flip-flops**

We will use the binary encoding ($s0 = 00$, $s1 = 01$, $s2 = 10$) and the number of flip-flops $= n = \log_2 3$. We will use two positive edge-sensitive D flip-flops.

4. **Reset strategy**

Let us use active low asynchronous reset input reset_n. For reset_n = 0, the sequence detector output is logic 0 and it holds default state S0. For the reset_n = 1, the sequence detector state transition happens on the rising edge of the clock.

5. **Let us document the entries in the state table to get state register logic**

The state table of the 101-sequence detector is shown Table 10.1.

So, we need to have two D flip-flops having asynchronous active low reset input reset_n.

6. **Use of the excitation table to design the next state logic**

The excitation table consists of the information about the present state, next state and excitation input (Table 10.2).

Table 10.1 State table of the 101 Mealy sequence detector

Input (data_in)	Present state ($q1\ q0$)	Next state ($q1^+q0^+$)
0	$s0$	$s0$
1	$s0$	$s1$
0	$s1$	$s2$
1	$s1$	$s1$
1	$s2$	$s1$
0	$s2$	$s0$

Table 10.2 Excitation table of the 101-sequence detector

Input (data_in)	Present state ($q1\ q0$)	Next state ($q1^{+}q0^{+}$)	Excitation input ($D1D0$)
0	$s0 = 00$	$s0 = 00$	$s0 = 00$
1	$s0 = 00$	$s1 = 01$	$s1 = 01$
0	$s1 = 01$	$s2 = 10$	$s2 = 10$
1	$s1 = 01$	$s1 = 01$	$s1 = 01$
1	$s2 = 10$	$s1 = 01$	$s1 = 01$
0	$s2 = 10$	$s0 = 00$	$s0 = 00$

Now let us use the K-map to deduce the Boolean equation for the next state logic (Figs. 10.3 and 10.4).

$$= \overline{\text{data\_in}}.q0$$

$$q0^{+} = D0 = \text{data\_in}$$

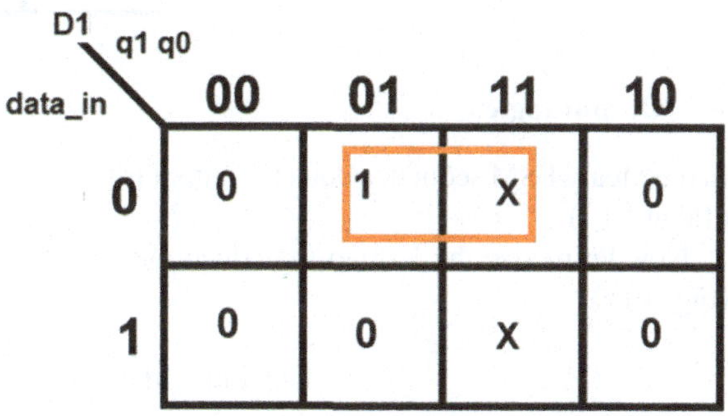

Fig. 10.3 K-map for the next state logic $D1$

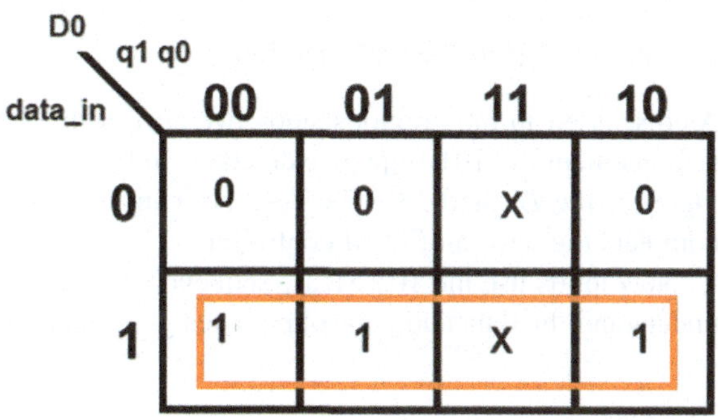

Fig. 10.4 K-map for the next state logic $D0$

Table 10.3 Truth-table of the output logic

Input (data_in)	Present state ($q1$ $q0$)	Next state ($q1^{+}q0^{+}$)	Output data_out
0	$s0 = 00$	$s0 = 00$	0
1	$s0 = 00$	$s1 = 01$	0
0	$s1 = 01$	$s2 = 10$	0
1	$s1 = 01$	$s1 = 01$	0
1	$s2 = 10$	$s1 = 01$	1
0	$s2 = 10$	$s0 = 00$	0

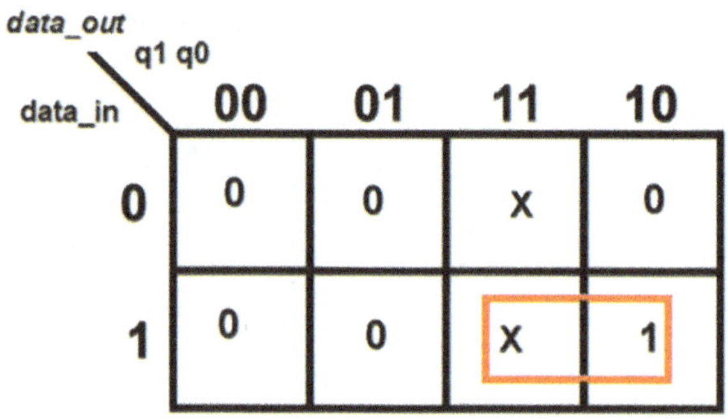

Fig. 10.5 K-map for the output logic

7. **Output logic**

In the Mealy FSM sequence detector output is function of the present state and inputs (Table 10.3).

Now let us use the K-map to deduce the Boolean equation for the output logic (Fig. 10.5).

$$\text{data\_out} = \text{data\_in}.q1$$

So, we need to have AND gate as an output logic.

8. **Let us sketch the FSM Design**

As discussed in the previous steps, we need to have two flip-flops and logic gates to implement the 101-sequence detector. In Fig. 10.6, the combo logic indicates the logic for the $D1$ and $D0$. The designer can use two flip-flops and combo logic to complete the design of FSM controller.

Now let us use the 101 Mealy sequence detector as control path logic and try to understand the data and control path design in much more detail.

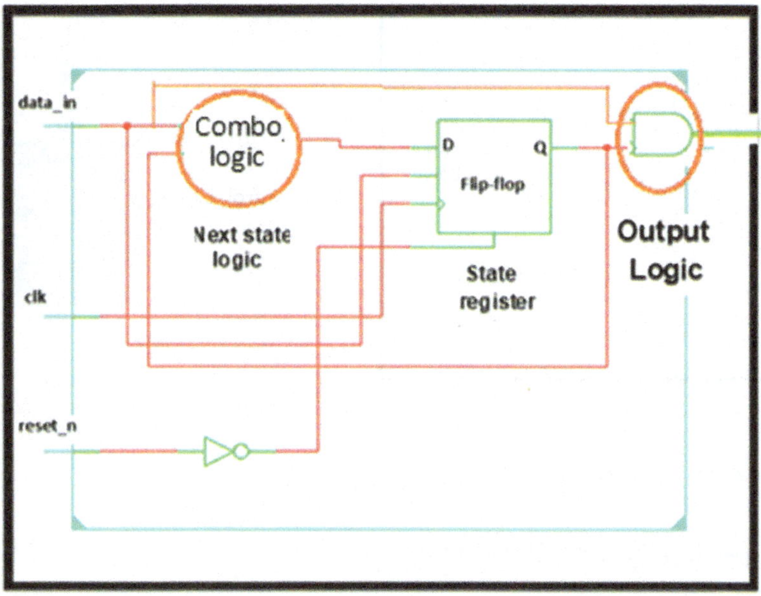

Fig. 10.6 Mealy sequence detector top-level logic to detect 101 overlapping sequence

10.4 Data and Control Path Design Techniques

Now let us use the FSM designed in the previous section to optimize the data and control paths. Most of the time we don't pay attention to have separate data and control paths, and this significantly affects on the area, speed and power of the design. As discussed in the previous section, if we have the better strategies to design the separate logic for the data and control path, then we can have significant improvement in the timing and even in the area.

Consider now the design scenario, the design requirement is to enable the data path logic after detecting the sequence 101 to transfer the 16-bit of the data from register A to output. In such scenario, we can use the following strategy.

1. **Design the control path logic**: Design the sequence detector to detect the sequence 101. The control logic uses the input as a clk, reset_n and data_in and generates a pulse at data_out if 101 sequence is detected.
2. **Design the data path logic**: In the data path let us have the register A which can hold the 16-bit data and when it is enabled it transfers the contents of A to output Y_out.

The design is shown in Fig. 10.7.

Now let us discuss the speed of the design and what are the different parameters used to find the maximum frequency for the design.

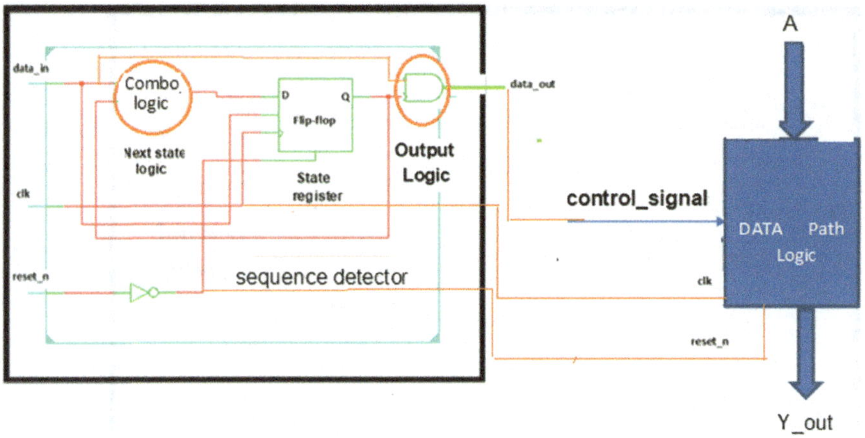

Fig. 10.7 Data and control path design strategy and use

10.5 Flip-Flop Timing Parameters

Important timing parameters of flip-flop are shown in Fig. 10.8, and they are:

1. Setup time (t_{su})
2. Hold time (t_h)
3. Propagation delay of flip-flop (t_{pff})

- **Setup Time (t_{su})**: The minimum amount of time for which the data input of the flip-flop should maintain the stable value prior to arrival of the active edge of the clock is called as setup time.

  During the setup time window if the data input changes, then the flip-flop output will be metastable which indicates the setup violation.
- **Hold time (t_h)**: The minimum amount of time for which the data input of the flip-flop should maintain the stable value after the arrival of the active edge of the clock is called as hold time.

Fig. 10.8 Timing parameters for D flip-flop

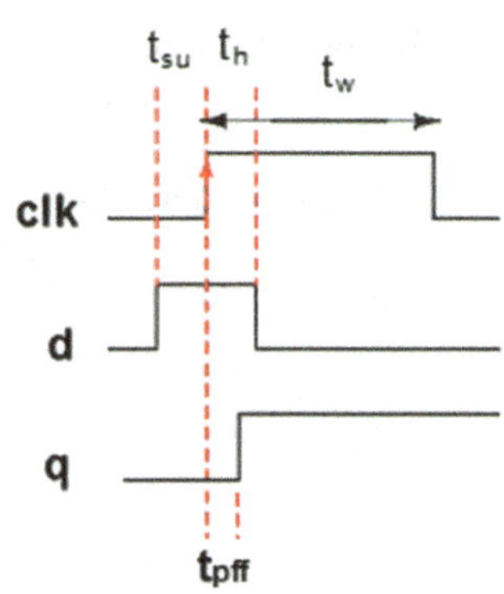

During the hold time window if the data input changes, then the flip-flop output will be metastable which indicates the hold violation.

- **Propagation Delay (tpdff = t_{cq})**: The amount of time required for the flip-flop to generate the valid output after the arrival of the active edge of the clock is called as propagation delay of flip-flop.

The propagation delay is also called as the **clock to q delay,** and it is also referred as $\mathbf{t_{cq}}$.

10.6 Example on Performance Improvement of the Design

Now let us discuss how we can use the timing parameters of the flip-flops and how we can get the maximum operating frequency for the design.

Consider the design (Fig. 10.9) of toggle flip-flop discussed in Chap. 7, consider $t_{\text{pff}} = 1$ ns, $t_{\text{combo}} = t_{\text{not}} =$ propagation delay of NOT gate $= 1$ ns and $t_{\text{su}} =$ setup time $= 1$ ns and $t_{\text{h}} =$ hold time $= 0.5$ ns. Let us find the maximum operating frequency for the design.

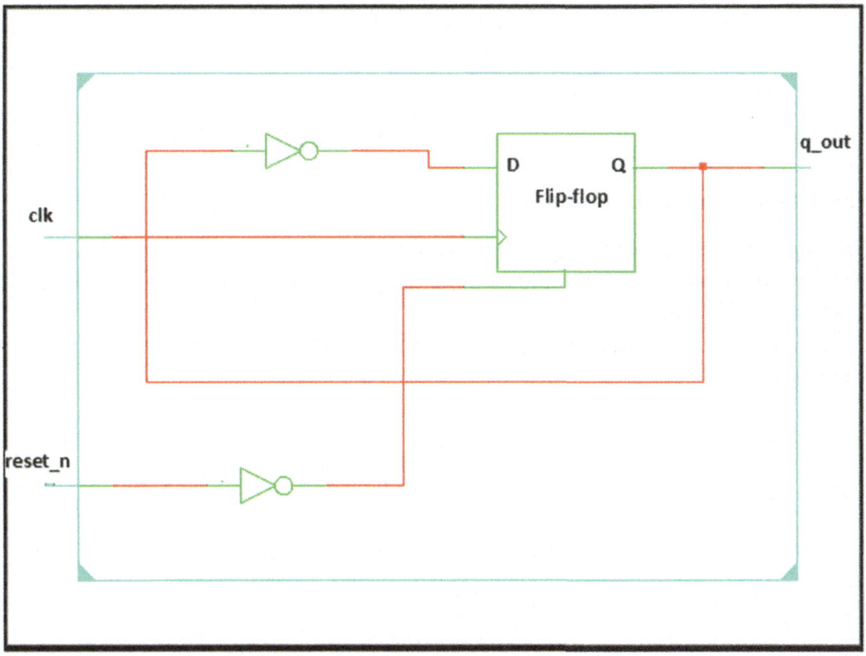

Fig. 10.9 Toggle flip-flop using D

To find the maximum operating frequency for the design, use the reg-to-reg timing path. In the design, the start point is clk and endpoint is D of the flip-flop.

1. Let us find out the data arrival time (AT). AT = $t_{pdff1} + t_{combo}$
2. The data required time is RT. RT = $T_{clk} - t_{su}$ because the data at D input should be stable by t_{su} margin prior to arrival of rising edge of the clock.
3. Now find setup slack. Slack = RT-AT and should be greater than or equal to 0.
4. Slack = RT-AT

$$= (T_{clk} - t_{su}) - (t_{pdff1} + t_{combo}).$$

5. If we equate the slack to zero, then we get

$$0 = (T_{clk} - t_{su}) - (t_{pdff1} + t_{combo})$$
$$T_{clk} = t_{pdff1} + t_{combo} + t_{su}$$

6. The maximum operating frequency of the design is f_{max}

$$\begin{aligned} f_{max} &= \frac{1}{T_{clk}} = \frac{1}{t_{pdff1} + t_{combo} + t_{su}} \\ &= \frac{1}{1\,\text{ns} + 1\,\text{ns} + 1\,\text{ns}} \\ &= \frac{1}{3\,\text{ns}} \\ &= 333.33\,\text{MHz} \end{aligned}$$

Now let us improve the design performance!

The logic in the design (Fig. 10.9) can be tweaked by removing NOT gate from the data path. Directly the complement of Q is given as data input to D flip-flop (Fig. 10.10).

This improves the area and speed for the design, and the maximum operating frequency we can get as

$$\begin{aligned} f_{max} &= \frac{1}{T_{clk}} = \frac{1}{t_{pdff1} + t_{su}} \\ &= \frac{1}{1\,\text{ns} + 1\,\text{ns}} \\ &= \frac{1}{2\,\text{ns}} \\ &= 500.00\,\text{MHz} \end{aligned}$$

We can use these techniques for the high-density designs also. For high-density designs, we can use the resource sharing and pipelining to improve the area and speed, respectively.

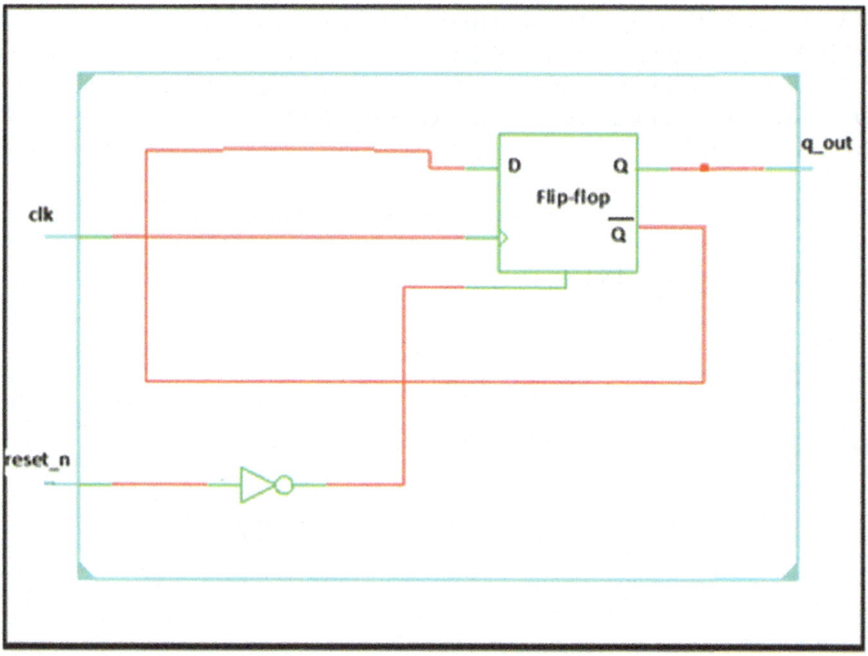

Fig. 10.10 Toggle D flip-flop

10.7 Exercises

Let us complete the exercises to find the maximum frequency for the design.

10.7.1 Exercise 1: Maximum Operating Frequency

Find the maximum operating frequency for the design shown in Fig. 10.11.

Solution: To find the maximum operating frequency for the design, use the reg-to-reg timing path. In the design, the start point is c1 and endpoint is D of second flip-flop.

Fig. 10.11 Reg-to-reg path

7. Let us find out the data arrival time (AT). AT $= t_{\text{pdff1}} + t_{\text{combo}}$
8. The data required time is RT. RT $= T_{\text{clk}} - t_{\text{su}}$ because the data at D input should be stable by t_{su} margin prior to arrival of rising edge of the clock.
9. Now find setup slack. Slack = RT-AT and should be greater than or equal to 0.
10. Slack = RT-AT

$$= (T_{\text{clk}} - t_{\text{su}}) - (t_{\text{pdff1}} + t_{\text{combo}})$$

11. If we equate the slack to zero, then we get

$$0 = (T_{\text{clk}} - t_{\text{su}}) - (t_{\text{pdff1}} + t_{\text{combo}})$$
$$T_{\text{clk}} = t_{\text{pdff1}} + t_{\text{combo}} + t_{\text{su}}$$

12. The maximum operating frequency of the design is $f_{\max}$

$$f_{\max} = \frac{1}{T_{\text{clk}}} = \frac{1}{t_{\text{pdff1}} + t_{\text{combo}} + t_{\text{su}}}.$$

10.7.2 Exercise 2: Timing Paths

Find the number of timing paths in the design shown in Fig. 10.12.

Solution: To find the number of timing paths for the design. Consider the start point as clock pin of the flip-flop, input port data_in and endpoint as data input of sequential element and data_out, data_out_1 that is output port.

1. **Input-to-reg path**: From data_in to D input of the first flip-flop.
2. **Reg-to-output path:** From the clock c2 to data_out
3. **Reg-to-reg path:** From clock c1 to D of the second flip-flop
4. **Input-to-output path:** From data_in to data_out_1. This is also called as combinational path.

So, the design has four timing paths.

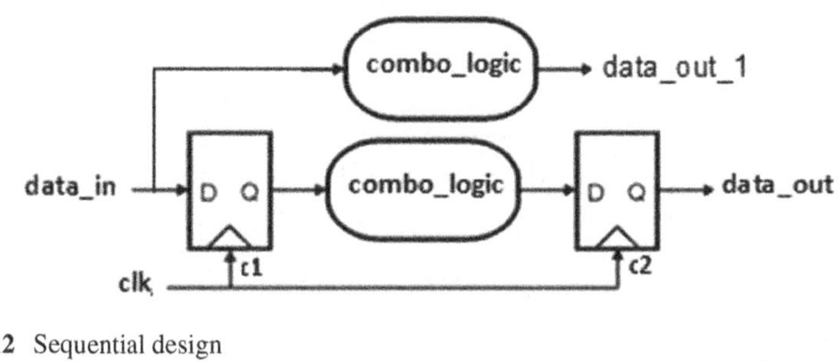

Fig. 10.12 Sequential design

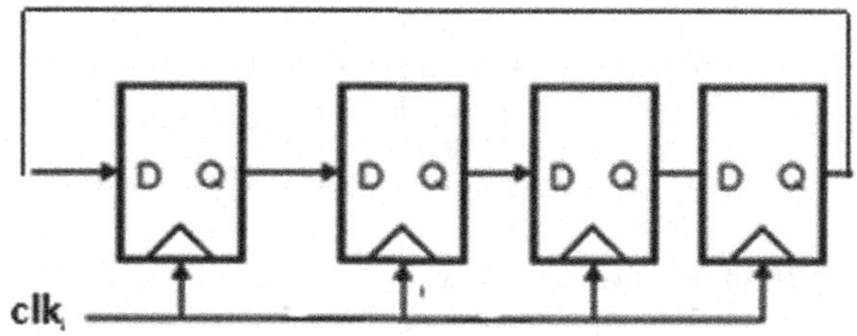

Fig. 10.13 Ring counter

10.7.3 Exercise 3: Maximum Operating Frequency

Find the maximum operating frequency for the design shown in Fig. 10.13.

Solution: To find the maximum operating frequency for the design, use any of the reg-to-reg timing path. In the design, the start point is clk and endpoint is D of second flip-flop.

1. Let us find out the data arrival time (AT). AT $= t_{pdff1}$
2. The data required time is RT. RT $= T_{clk} - t_{su}$ because the data at D input should be stable by t_{su} margin prior to arrival of rising edge of the clock.
3. Now find setup slack. Slack = RT-AT and should be greater than or equal to 0.
4. Slack = RT-AT

$$= (T_{clk} - t_{su}) - (t_{pdff1})$$

5. If we equate the slack to zero, then we get

$$0 = (T_{clk} - t_{su}) - (t_{pdff1})$$
$$T_{clk} = t_{pdff1} + t_{su}$$

6. The maximum operating frequency of the design is f_{max}

$$f_{max} = \frac{1}{T_{clk}} = \frac{1}{t_{pdff1} + t_{su}}$$

10.7.4 Exercise 4: Positive Clock Skew and Maximum Operating Frequency for the Design

Find the maximum operating frequency for the design shown in Fig. 10.14.

Solution: To find the maximum operating frequency for the design use the reg-to-reg timing path. In the design, the start point is c1 and endpoint is D of second flip-flop.

1. Let us find out the data arrival time (AT). AT $= t_{pdff1} + t_{combo}$

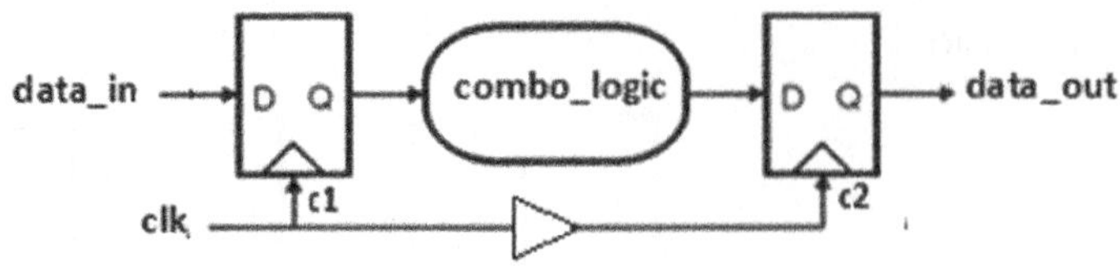

Fig. 10.14 Positive clock skew

2. The data required time is RT. RT = $T_{clk} - t_{su} + t_{buf}$ because the data at D input should be stable by t_{su} margin prior to arrival of rising edge of the clock and clock is delayed by the buffer delay.
3. Now find setup slack. Slack = RT-AT and should be greater than or equal to 0.
4. Slack = RT-AT

$$= (T_{clk} - t_{su} + t_{buf}) - (t_{pdff1} + t_{combo})$$

5. If we equate the slack to zero, then we get

$$0 = (T_{clk} - t_{su} + t_{buf}) - (t_{pdff1}) - (t_{combo})$$
$$T_{clk} = t_{pdff1} + (t_{combo}) + t_{su} - t_{buf}$$

6. The maximum operating frequency of the design is f_{max}

$$f_{max} = \frac{1}{T_{clk}} = \frac{1}{t_{pdff1} + t_{combo} + t_{su} - t_{buf}}$$

10.7.5 Exercise 5: Negative Clock Skew and Maximum Operating Frequency for the Design

Find the maximum operating frequency for the design shown in Fig. 10.15.

Solution: To find the maximum operating frequency for the design, use the reg-to-reg timing path. In the design, the start point is c1 and endpoint is D of second flip-flop.

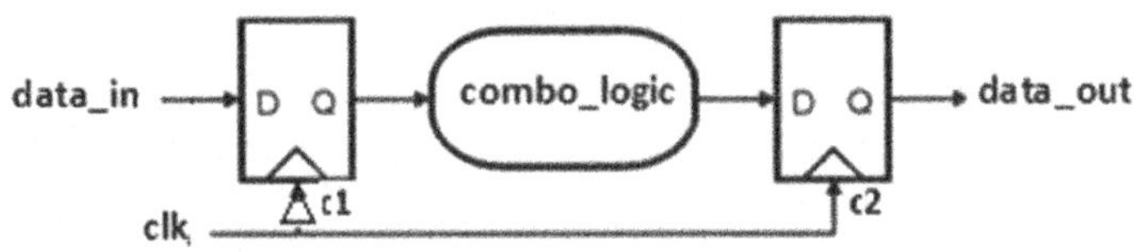

Fig. 10.15 Negative clock skew

1. Let us find out the data arrival time (AT). AT $= t_{\text{pdff1}} + t_{\text{combo}}$
2. The data required time is RT. RT $= T_{\text{clk}} - t_{\text{su}} - t_{\text{buf}}$ because the data at D input should be stable by t_{su} margin prior to arrival of rising edge of the clock and clock is delayed by the buffer at flip-flop 1.
3. Now find setup slack. Slack = RT-AT and should be greater than or equal to 0.
4. Slack = RT-AT

$$= (T_{\text{clk}} - t_{\text{su}} - t_{\text{buf}}) - (t_{\text{pdff1}} + t_{\text{combo}})$$

5. If we equate the slack to zero, then we get

$$0 = (T_{\text{clk}} - t_{\text{su}} - t_{\text{buf}}) - (t_{\text{pdff1}} + t_{\text{combo}})$$
$$T_{\text{clk}} = t_{\text{pdff1}} + t_{\text{combo}} + t_{\text{su}} + t_{\text{buf}}$$

6. The maximum operating frequency of the design is $f_{\max}$

$$f_{\max} = \frac{1}{T_{\text{clk}}} = \frac{1}{t_{\text{pdff1}} + t_{\text{combo}} + t_{\text{su}} + t_{\text{buf}}}$$

10.8 Important Takeaways

Following are few of the important points to conclude this chapter.

1. The FSM should have better data and control path design.
2. The goal is to improve the design speed by minimizing the combinational delay in the reg-to-reg path.
3. The flip-flop setup time is minimum amount of time during which data should be stable prior to arrival of the active edge of the clock.
4. The flip-flop hold time is minimum amount of time during which data should be stable after the arrival of the active edge of the clock.
5. The flip-flop propagation delay is the amount of time required to get output data after arrival of the active edge of the clock.
6. The start point is clk port and input ports.
7. The endpoint is output ports, data input of the D flip-flop.
8. The maximum operating frequency is dependent on the timing parameters of the flip-flop and the combinational delay.

Chapter 11
Advanced Design Techniques-2

The various efficient design techniques are useful during the architecture and micro-architecture design..

In the previous chapter, we have discussed the advanced digital design techniques. In this chapter, we will focus on the architecture design for the given functional specifications. The chapter is even useful to understand the design-specific scenarios like multiple clock domains, multiple power domains, synchronizers, design specific scenarios and the performance improvement for the design.

11.1 Multiple Clock Domain Designs

Most of the time, we have the multiple clock domain designs. Consider the design which has the general-purpose processor, video encoder/decoder and memory controllers. The general-purpose processor operates on clk1 = 500 MHz, the video encoder/decoder operates on the clk2 = 250 MHz, and the memory controller operates at the frequency of clk3 = 333.33 MHz. So, we need to have three different clock sources, and such type of the design is called as the multiple clock domain design (Fig. 11.1).

What are issues in such designs?

In the multiple clock domain designs, the major issue is exchange of the data between the clock domains. There is issue of the data convergence and can be overcome by using the control and data path synchronizers.

We can think of using the level synchronizers in the control path and FIFO synchronizers in the data path. The following section discusses the issues in the multiple clock domain designs!

V. Taraate, *Digital Design Techniques and Exercises*,
https://doi.org/10.1007/978-981-16-5955-3_11

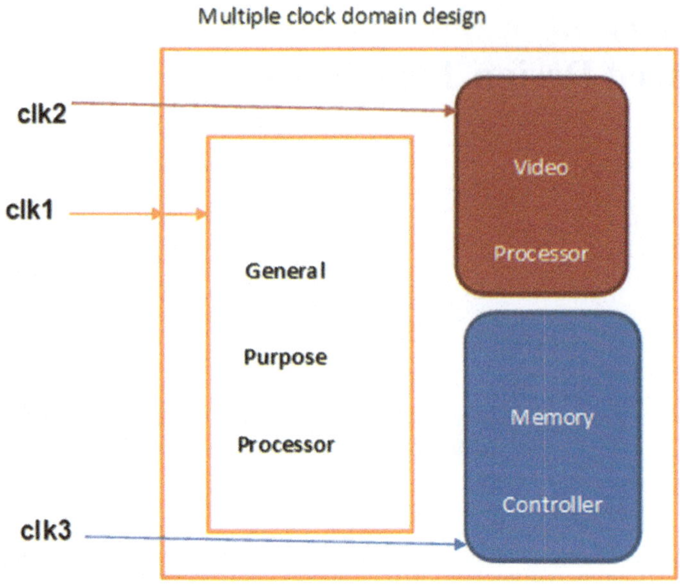

Fig. 11.1 Multiple clock domain design

11.2 Metastability

Consider the design shown in Fig. 11.2. The data_in is input of the design and sampled on the rising edge of the clock. If the other design which works at different clock frequency drives the data_in, then the design has multiple clock boundaries. In such scenario due to phase difference between the multiple clocks, the first flip-flop (Fig. 11.2) goes into the metastable state. The meta_data indicates the flip-flop output is metastable, and hence there is timing violation for the first flip-flop.

Themetastability indicates the data output is not valid, and to get the valid data output, the design needs to use the multi-flop-level synchronizers.

The timing sequence is shown in Fig. 11.3. As shown the output of the first flip-flop is in the metastable state, and the data_out output from the output flip-flop is having the valid legal state.

Fig. 11.2 Level synchronization concept

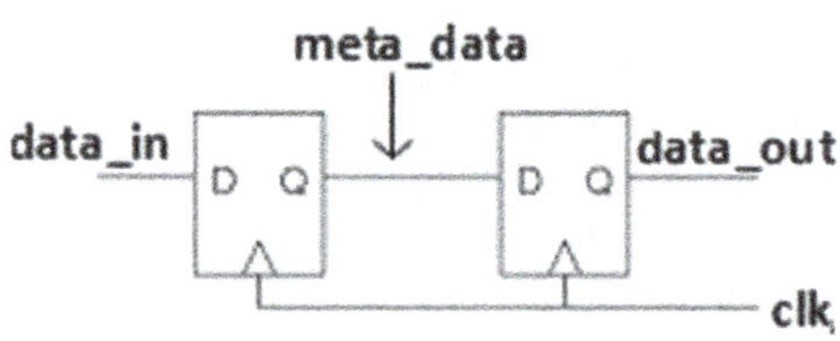

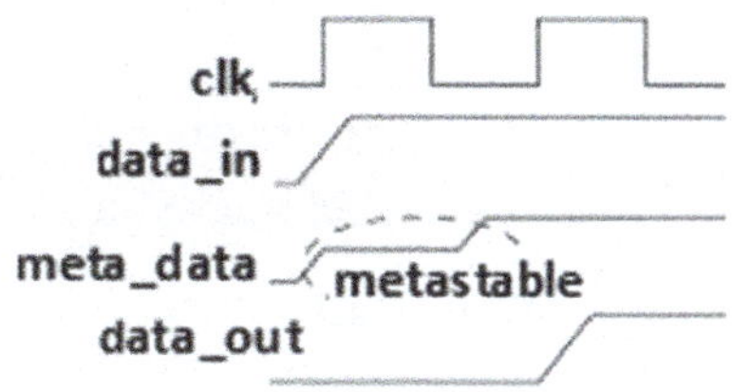

Fig. 11.3 Timing sequence for Fig. 11.2

11.3 Control Path Synchronizer

As discussed in the previous section, we can think of using the level synchronizers in the control path. For the level synchronizer, the first flip-flop output is metastable and should be ignored by setting the false path in the timing analysis. The control path synchronizer which uses two flip-flops in cascade is shown in Fig. 11.4.

There are various other synchronization techniques to pass the signal between the multiple clock domains. Few of them are mux synchronizers, pulse synchronizers.

Consider the use of the level synchronizer in the control path. Consider the clock domain 1 operates at 100 MHz and clock domain 2 at 50 MHz. To pass the control signals from one of the clock domain to another clock domain, we can think of use of the level synchronizers (Fig. 11.5).

Fig. 11.4 Two-flop-level synchronizer

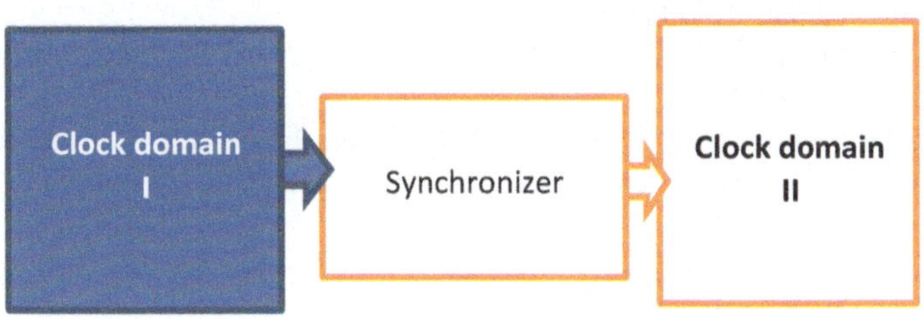

Fig. 11.5 Passing the control signal between the clock domain

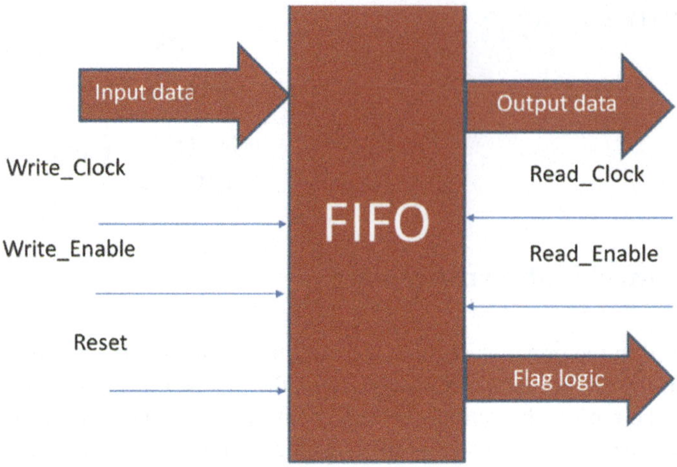

Fig. 11.6 FIFO in the data path

11.4 Data Path Synchronizer

We can use the asynchronous FIFO to pass the burst of data between the clock domains.

The FIFO has input and output data. The write clock and read clock domain. The basic architecture of the FIFO is shown in Fig. 11.6.

The FIFO is used in the data path to transfer the burst of the data. The write clock domain and read clock domain frequencies are different, and during the design, we need to understand about the depth of FIFO required. The FIFO depth calculation is discussed in the exercises.

Depending on the write clock frequency the burst of the data is written into the FIFO memory when FIFO is not full.

When FIFO is not empty, then depending on the read clock frequency the data can be read.

During the design of the data path synchronizer, the objective of the designer is to design the FIFO logic to achieve the data convergence.

11.5 Multiple Power Domain Designs

The way in which we have the multiple clocks in the design we can have the multiple power domains. That is the different design units operates on the various voltage levels. For example, processor operates at 1.8 V, the video controller operates at 3.3 V, and the memory controller operates at 2.5 V (Fig. 11.7).

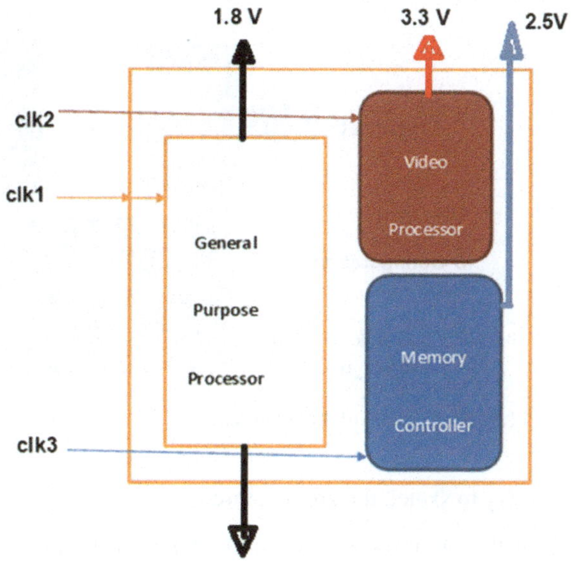

Fig. 11.7 Multiple clock and power domain design

As shown in Fig. 11.7, we have three power domains in the design. For such kind of designs, we need to have better power domain management! We can use the following strategies while designing the multiple power domain designs.

1. Have the power-up sequence using the state machine
2. Use the level shifters and isolation cells while communicating between multiple power domains.
3. Use the retention cells to retain the state of the block during power shut-down.
4. Have the low-power design architecture using the low-power cells.

Most of the design architectures demands the use of the high-speed techniques and low-power concepts. So let us discuss use of these concepts to design the architecture for the given specifications.

11.6 Architecture-Level Designs

Consider the design of the high-speed data transfer logic between the processor and the IO. So let us discuss how we can design the architecture and micro-architecture for such kind of designs.

1. Understand about the speed that is operating frequency of the processor.
2. Understand about the operating frequency of the IO device.
3. The goal is to design the IO controller where the processor communicates.
4. So, try to find out the speed and data rate of the channel to transfer or to accept the burst of data.

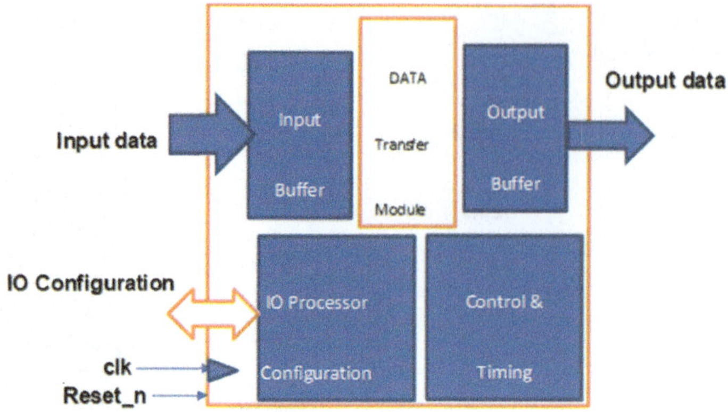

Fig. 11.8 Architecture of the IO processor

5. Try to sketch the architecture.

At the top-level we can think about the IO processor which can be interfaced between the main processor and IO devices to transfer the data.

From the design specifications we can design the architecture as shown in the figure. As shown in Fig. 11.8, the main functional blocks to transfer the data between the main processor and IO are

1. Input buffer
2. Output buffer
3. Data transfer module
4. IO configuration module
5. Control and timing unit.

For each module, we need to identify the interface and control signals. We should create the micro-architecture for each block if the goal is to use this for RTL design. For the system design, we should have different thought process and will be discussed in the next chapter.

11.7 How We Can Improve the Design Performance

To improve the design performance at the architecture level, we need to understand about the top-level design constraints such as area, speed and power.

Following are few of the strategies we can adopt to improve the design performance.

1. Use the better partitioning strategies by using separate functional blocks.
2. Have the parallelism in the design to perform multiple operations at a time. For example, fetch of the data from various devices at a time. This may increase the area but improves the speed of the design (Fig. 11.9).

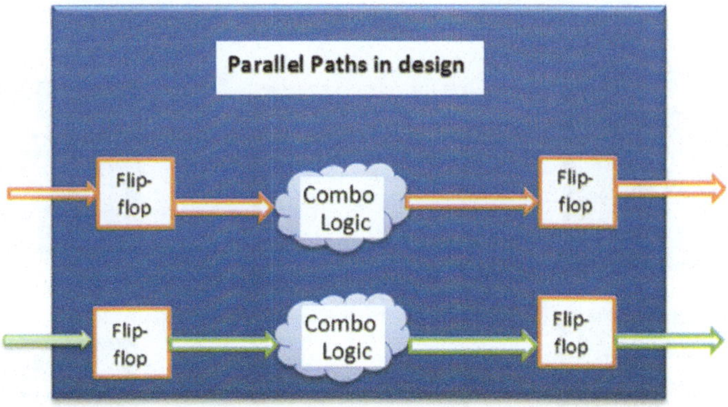

Fig. 11.9 Parallel paths in the design

3. Have the clock distribution to achieve the better timing.
4. Use the reset logic to initialize the functional blocks at power-on.
5. Avoid the cascade of the functional blocks to improve the timing.
6. Use the pipelining and resource sharing while developing the processing logic. For example, fetch, decode, execute and store can be performed by having the 4-stage pipelined architecture.
7. Isolate the power domains and clock domains using the design strategies.
8. Have the design policies for the memory buffers and the internal storage.
9. Use the high-speed interfaces to minimize the latency.
10. Use the glitch suppressing circuits and mechanisms in the design.
11. Use the clock gating cells to reduce the dynamic power dissipation (Fig. 11.10).

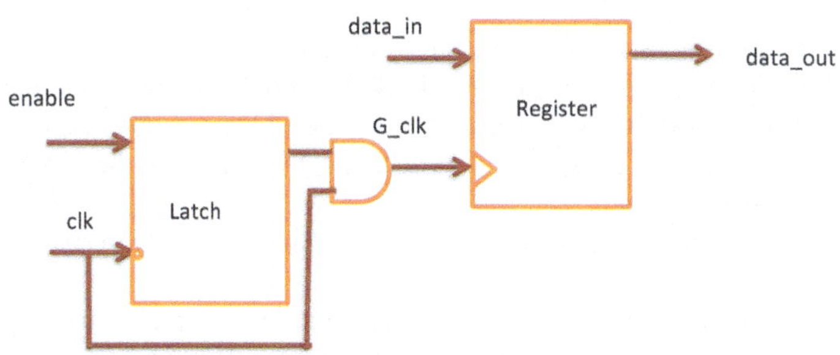

Fig. 11.10 Low-power clock gating cell

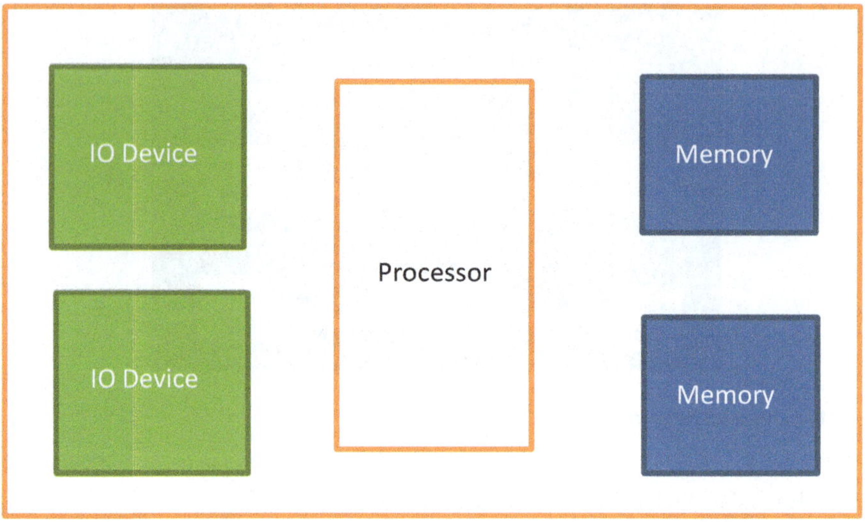

Fig. 11.11 Basic components in the system design

11.8 The Digital Systems and Design

Consider the design of the digital system. As most of us are familiar with the fact that the digital system consists of the

1. Processor
2. Input–output devices
3. Memories

Processor controls the data transfer between the IO and memory devices depending on the IO or memory instructions. The processor has address, data and control buses. The address bus is used to transfer the address of IO or memory device. The control bus generates the read and write control signals, and the data bus is used to transfer or accept the data.

The memory devices can be ROM or RAM and can be interfaced with the processor using the desired decoding schemes.

The IO devices like keyboard, display, ADC and DAC are used to communicate with the processor using the desired decoding schemes (Fig. 11.11).

With the above components the system has the power supply, clock generation logic, reset generation logic and other high speed data transfer mechanisms.

11.9 Exercises

Let us complete few exercises on the design techniques discussed in this chapter.

11.9.1 Exercise 1: FIFO Depth Calculation

Consider the design having write clk frequency $f_1 = 100$ MHz and read clock frequency $f_2 = 50$ MHz, and the burst length is 40 byte; find the depth of FIFO?

Solution: To find the depth of the FIFO, use the following steps.

1. **Time required to write single data byte** (T_w)

$$T_w = 1/100\,\text{MHz} = 10\,\text{ns}$$

2. **Time required to write burst of the data that is 40 bytes** (T_{b_w})

$$T_{b\_w} = T_w * \text{Burst length} = 10\,\text{ns} * 40 = 400\,\text{ns}$$

3. **Time required to read one data** (T_r)

$$T_r = 1/50\,\text{MHz} = 20\,\text{ns}$$

4. **The number of data reads in duration of** T_{b_w}

$$\text{No of reads} = 400\,\text{ns}/20\,\text{ns} = 20$$

5. **The depth of FIFO**

$$\text{Depth of FIFO} = \text{Burst length} - \text{No of reads} = 40 - 20 = 20.$$

11.9.2 Exercise 2: FIFO Depth Calculation

Consider the design having write clk frequency $f_1 = 100$ MHz and read clock frequency $f_2 = 50$ MHz, and the burst length is 40 bytes. The number of idle cycles between two writes is 1, and number of idle cycle between two reads is 3. Find the depth of FIFO?

Solution: To find the depth of the FIFO, use the following steps.

1. **Time required to write single data byte** (T_w)

   One idle cycle between two writes so for two cycles one data is written.

$$T_w = 2 * (1/100\,\text{MHz}) = 20\,\text{ns}$$

2. **Time required to write burst of the data that is 40 bytes** (T_{b_w})

$$T_{b\_w} = T_w * \text{Burst length} = 20\,\text{ns} * 40 = 800\,\text{ns}$$

3. **Time required to read one data (T_r)**
   For two successive reads three cycles (2 idle + 1 = 3 cycles)

$$T_r = 4 * (1/50\,\text{MHz}) = 80\,\text{ns}$$

4. **The number of data reads in duration of T_{b_w}**

$$\text{No of reads} = 800\,\text{ns}/80\,\text{ns} = 10$$

5. **The depth of FIFO**

$$\text{Depth of FIFO} = \text{Burst length} - \text{No of reads} = 40 - 10 = 30.$$

11.9.3 Exercise 3: FIFO Depth Calculation

Consider the design having write clk frequency $f_1 = 50$ MHz and read clock frequency $f_2 = 100$ MHz, and the burst length is 100 bytes. The number of idle cycles between two writes is 1, and number of idle cycle between two reads is 3. Find the depth of FIFO?

Solution: To find the depth of the FIFO, use the following steps.

1. **Time required to write single data byte (T_w)**

   One idle cycle between two writes so for two cycles one data is written.

$$T_w = 2 * (1/50\,\text{MHz}) = 40\,\text{ns}$$

2. **Time required to write burst of the data that is 100 bytes (T_{b_w})**

$$T_{b\_w} = T_w * \text{Burst length} = 40\,\text{ns} * 100 = 4000\,\text{ns}$$

3. **Time required to read one data (T_r)**

   For two successive reads three cycles (2 idle + 1 = 3 cycles)

$$T_r = 4 * (1/100\text{MHz}) = 40\,\text{ns}$$

4. **The number of data reads in duration of T_{b_w}**

$$\text{No of reads} = 4000\,\text{ns}/40\,\text{ns} = 100$$

5. **The depth of FIFO**

$$\text{Depth of FIFO} = \text{Burst length} - \text{No of reads} = 100 - 100 = 0\ \text{NO FIFO required.}$$

11.9.4 Exercise 4: FIFO Depth Calculation

Consider the design having write clk frequency $f_1 = 25$ MHz and read clock frequency $f_2 = 40$ MHz, and the burst length is 100 bytes. The number of idle cycles between two writes is 1, and number of idle cycle between two reads is 3. Find the depth of FIFO?

Solution: To find the depth of the FIFO, use the following steps.

1. **Time required to write single data byte** (T_w)

   One idle cycle between two writes so for two cycles one data is written.

$$T_w = 2 * (1/25\,\text{MHz}) = 80\,\text{ns}$$

2. **Time required to write burst of the data that is 100 bytes** (T_{b_w})

$$T_{b\_w} = T_w * \text{Burst length} = 80\,\text{ns} * 100 = 8000\,\text{ns}$$

3. **Time required to read one data** (T_r)

   For two successive reads three cycles (2 idle + 1 = 3 cycles)

$$T_r = 4 * (1/40\text{MHz}) = 100\,\text{ns}$$

4. **The number of data reads in duration of** T_{b_w}

$$\text{No of reads} = 8000\,\text{ns}/100\,\text{ns} = 80$$

5. **The depth of FIFO**

$$\text{Depth of FIFO} = \text{Burst length} - \text{No of reads} = 100 - 80 = 20.$$

11.9.5 Exercise 5: FIFO Depth Calculation

Consider the design having write clk frequency $f_1 = 50$ MHz and read clock frequency $f_2 = 50$ MHz, and the burst length is 80 bytes. The number of idle cycles between two writes is 1, and number of idle cycle between two reads is 3. Find the depth of FIFO?

Solution: To find the depth of the FIFO, use the following steps.

1. **Time required to write single data byte** (T_w)

   One idle cycle between two writes so for two cycles one data is written.

$$T_w = 2 * (1/50\,\text{MHz}) = 40\,\text{ns}$$

2. **Time required to write burst of the data that is 80 bytes (T_{b_w})**

$$T_{b\_w} = T_w * \text{Burst length} = 40\,\text{ns} * 80 = 3200\,\text{ns}$$

3. **Time required to read one data (T_r)**
   For two successive reads three cycles (2 idle + 1 = 3 cycles)

$$T_r = 4 * (1/50\text{MHz}) = 80\,\text{ns}$$

4. **The number of data reads in duration of T_{b_w}**

$$\text{No of reads} = 3200\,\text{ns}/80\,\text{ns} = 40$$

5. **The depth of FIFO**

$$\text{Depth of FIFO} = \text{Burst length} - \text{No of reads} = 80 - 40 = 40.$$

11.10 Important Takeaways

Following are few of the important points to conclude this chapter.

1. For the multiple clock domain designs, use the level synchronizers in the control path
2. Use the FIFO synchronizers to pass the data between the clock domain.
3. For multiple power domain designs, use the level shifters, isolation cells and retention cells.
4. Use the FIFO depth calculations to find the depth of the FIFO in the multiple clock domain designs.
5. Have the provision of the parallelism in the architecture and micro-architecture designs.
6. Have the decoding strategies for the IO and memory decoders in the system design.

Chapter 12
System Design and Considerations

For the digital system design, we should have high speed processor, IO devices and Memory devices as basic components.

In previous chapters, we have discussed various digital design techniques and exercises. Most of the time, we experience the need of the digital design techniques to design the digital systems. If we consider any digital system, then the understanding of the digital design techniques and their use is helpful to the engineers to design and implement the systems. The main considerations are the area, speed and power requirements for these systems and their efficient understanding while implementing the digital systems. In this context, the chapter discusses the use of the digital design techniques in the system design and other important goals.

12.1 System Design

If we consider any system design, then as discussed in the previous chapter we should have the

1. Processor
2. Input output devices
3. Memories.

Figure 12.1 is the top-level component understanding of the digital system. Apart from the components shown, we need to have the clock, power supply and reset logic. Even the system can have additional high-speed interfaces and other test and debug module on board. The objective of the chapter is to understand the digital system and how these components are interfaced to establish the communication.

V. Taraate, *Digital Design Techniques and Exercises*,
https://doi.org/10.1007/978-981-16-5955-3_12

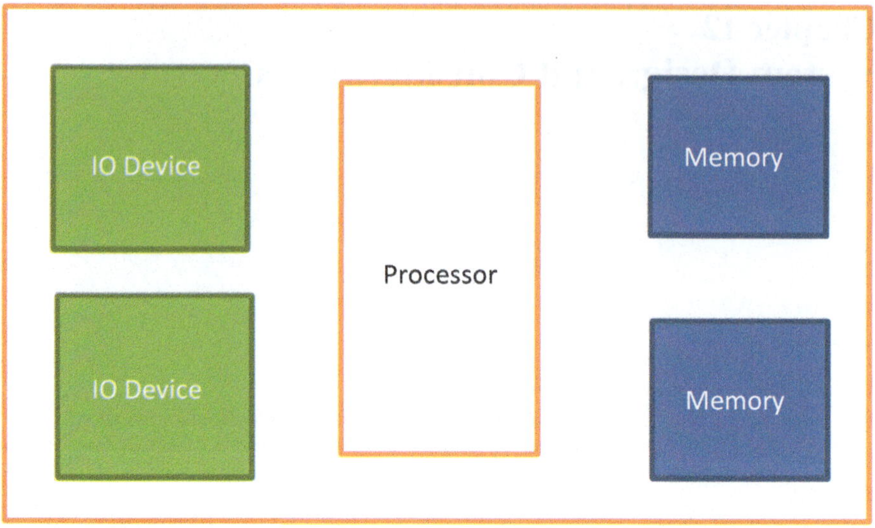

Fig. 12.1 System design components

12.2 What We Need to Think About?

While developing the system, we need to think about the following important points.

1. The application and use of the system
2. Identify the components
3. IO devices: Type of IO devices
4. Memory devices: Types of memory needed in the system
5. How to identify the devices?
6. What should be decoding logic?
7. What is the speed
8. Compatibility considerations
9. What are the area, speed and power requirements?
10. Supply voltages and power supply design
11. Clock and reset management
12. Board design and the interface issue.

By considering all the above points, we can design the digital system. We can use the microprocessor, microcontroller or FPGA as processing unit.

12.3 Important Considerations

To design the digital system, with the least propagation delay and the lower power dissipation is the goal of the system designer! Even we need to consider about the

1. Compatibility of the devices used
2. Fanout
3. Noise Margin.

Let us discuss then with some design example.

1. **Compatibility**: Two devices are compatible it indicates that their logic levels are same. They can be directly interfaced with each other. As shown the output of AND drives the NOT gate directly that means they are compatible with each other (Fig. 12.2).

2. **Fanout**: The maximum load that can be driven by the driver is called as fanout. While designing any digital system, we need to understand the fanout of the design (Fig. 12.3).

3. **Noise Margin**: The maximum acceptable noise in the digital system is important parameter and treated as noise margin.

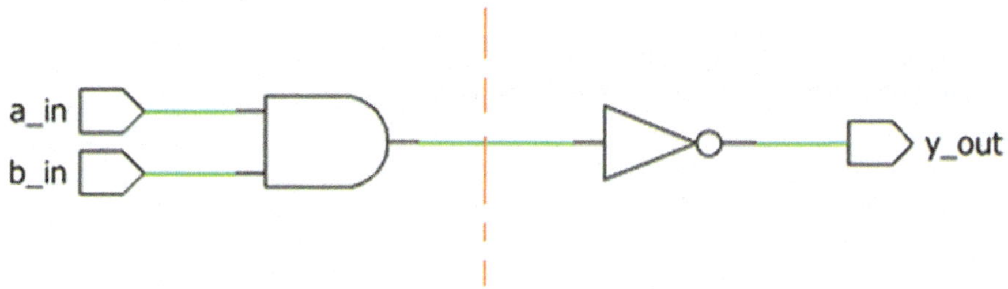

Fig. 12.2 Compatible devices

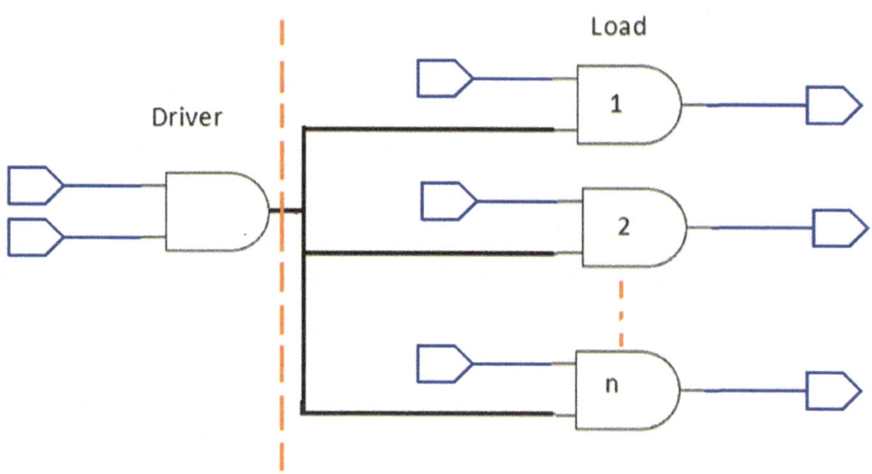

Fig. 12.3 Fanout of the design

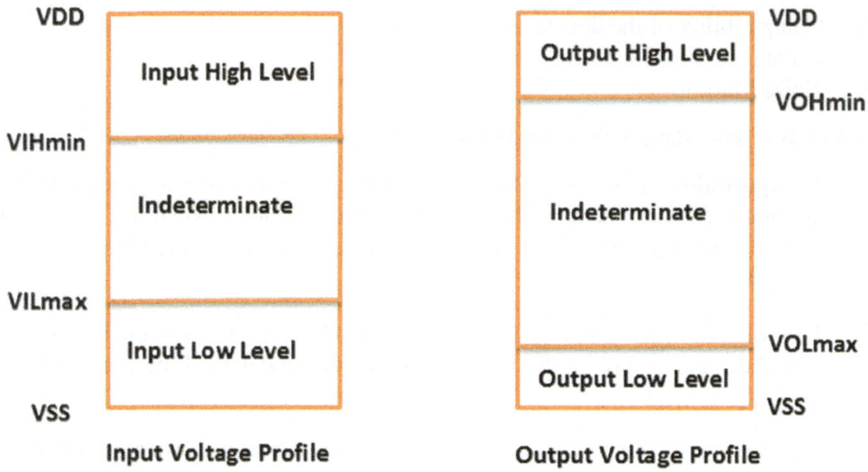

Fig. 12.4 Voltage profiles and noise margin

Consider the voltage profiles shown in Fig. 12.4.

The noise margin is V_n and it is defined as

$$V_n = V_{\mathrm{IL\,max}} - V_{\mathrm{OL\,max}}$$

where

V_{ILmax} Maximum low-level input voltage
V_{OLmax} Maximum low-level output voltage

The noise margin is V_n is also defined as

$$V_n = V_{\mathrm{OH\,min}} - V_{\mathrm{IH\,min}}$$

where

V_{IHmin} Minimum high-level input voltage
V_{OHmin} Minimum high-level output voltage

All the above parameters are useful during selection of the various system components.

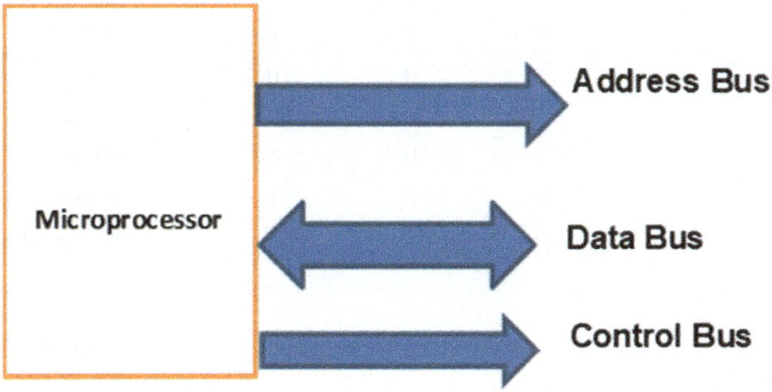

Fig. 12.5 System bus

12.4 Let Us Understand the Microprocessor Capabilities

The microprocessor should perform the data transfer, arithmetic, logical and branching operations. The microprocessor should have the dedicated system bus to carry address, data and control signal.

Now for better understanding, consider the 4-bit microprocessor which has the 4-bit address bus, 8-bit data bus and control signal as RD, WR, IO_M to control the data transfer between the memory/IO and processor.

Address Bus: The bus is used to carry the address of memory or IO device. Consider A0 to A3 as unidirectional address lines from the processor.

Data Bus: The bidirectional bus to carry the data. The data can be exchanged between the processor and IO/memory devices. Consider DO-D7 as bidirectional data bus.

Control Bus: Used to control the read and write for the IO or memory.

Figure 12.5 gives information about the various buses used in the system design.

12.5 Control Signal Generation Logic

Let us use the understanding of the decoders and combinational logic to generate the control signals for the memory and IO devices.

As discussed about the main control signals are RD, WR, IO_M and used to perform the read and write to/from IO/memory devices (Table 12.1).

To design the control logic, let us use the entries documented in Table 12.1. We can use the 3:8 decoder, and we can generate the control signal for the memory/IO read/write.

As documented when RD = 1, then it performs the memory read if IO_M = 0 and IO read when IO_M = 1.

Table 12.1 Control signals

RD	WR	IO_M	Operation
1	0	0	Memory read
0	1	0	Memory write
1	0	1	IO read
0	1	1	IO write
0	0	X	No operation

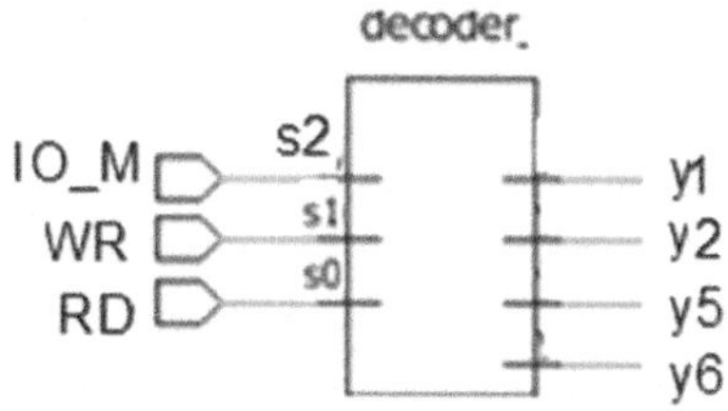

Fig. 12.6 Memory/IO control signal logic

As documented when WR = 1, then it performs the memory write if IO_M = 0 and IO write when IO_M = 1.

When both the RD and WR are logic 0, then the processor will not perform any operation (Fig. 12.6).

The control signals are generated by using the 3:8 decoder. To get memory read I0_M = 0, WR = 0, RD = 1 hence as shown the y1 = 1.

To get memory write I0_M = 0, WR = 1, RD = 0 hence as shown the y2 = 1.

To get IO read I0_M = 1, WR = 0, RD = 1 hence as shown the y5 = 1.

To get IO write I0_M = 1, WR = 1, RD = 0 hence as shown the y6 = 1.

At a time, only one output of decoder is active high.

12.6 IO Devices and Communication with the Processor

Let us consider the read and write transactions for the IO devices. How we can establish the communication with the IO devices that is particularly important point to discuss!

Consider that, the system requirement is of 4, IO devices and each IO device has four ports.

So, what should be our strategy?

To identify one of the port let us use two address line. Why? Because 4 ports = 2^2. Power of 2 indicates the number of address lines needed. Table 12.2 gives information about the identification of one of the port depending on the status of $A1$, $A0$.

Table 12.2 IO ports and identification

en = IO_M	*A*1	*A*0	Description
1	0	0	IO device port 0 (y0) selected
1	0	1	IO device port 1 (y1) selected
1	1	0	IO device port 2(y2) selected
1	1	1	IO device port 3 (y3) selected
0	X	X	None of the IO device selected

Table 12.3 IO device selection

en = IO_M	*A*3	*A*2	Description
1	0	0	IO device #1 selected
1	0	1	IO device #2 selected
1	1	0	IO device #3 selected
1	1	1	IO device #4 selected
0	X	X	None of the IO device selected

Now each IO device has specific address. The IO #1 has address 0000-0011, the IO #2 has address 0100-0111, the IO #3 has address 1000-1011, and the IO #4 has address 1100-1111. So let us use the address lines *A*3, *A*2 and select one of the IO device (Table 12.3).

Now let us interface the four IO devices with the microprocessor. As shown in Fig. 12.7, to enable the IO decoder the en = IO_M (when 1 IO decoder is enabled) is used.

The address lines *A*1, *A*0 are used directly to identify the IO port.

The address lines *A*3, *A*2 are used to select one of the IO device at a time using the 2:4 decoder.

12.7 Memory Devices and Communication with the Processor

Let us consider the read and write transactions for the memory devices. How we can establish the communication with the memory devices that is discussed in this section!

Consider that the system requirement is of four memory devices, and each memory device has four, 8-bit registers.

So, what should be our strategy?

To identify one of the memory device, let us use two address line. Why? Because four memory locations/registers = 2^2. Power of 2 indicates the number of address lines needed. Table 12.4 gives information about the identification of one of the memory location depending on the status of *A*1, *A*0.

Fig. 12.7 IO interfacing

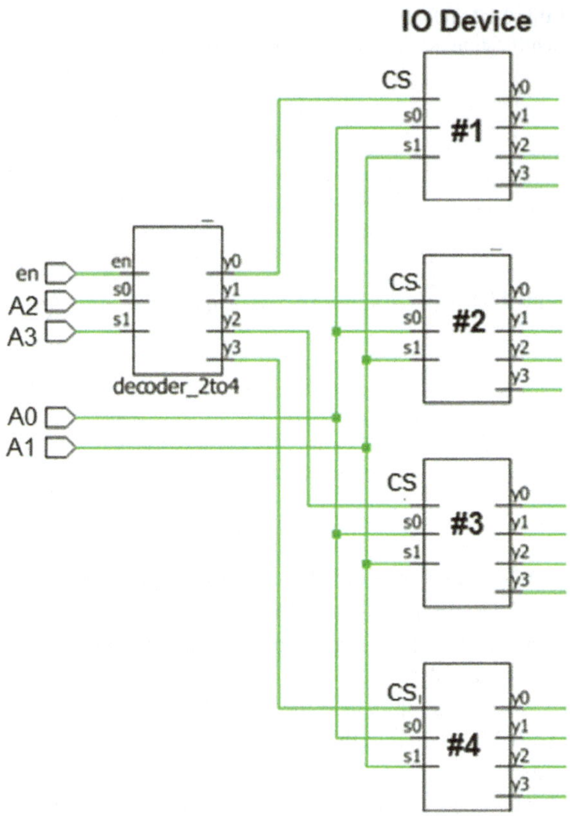

Table 12.4 Memory devices and identification

en = IO_M	*A*1	*A*0	Description
0	0	0	Memory device location 0 selected
0	0	1	Memory device location 1 selected
0	1	0	Memory device location 2 selected
0	1	1	Memory device location 3 selected
1	X	X	None of the memory device selected

Now each memory device has specific address. The memory #1 has address 0000-0011, the memory #2 has address 0100-0111, the memory #3 has address 1000-1011, the memory #4 has address 1100-1111. So let us use the address lines *A*3, *A*2 and select one of the memory device (Table 12.3).

Now let us interface the four memory devices with the microprocessor. As shown in Fig. 12.8, to enable the memory decoder the en = IO_M (when 0 memory decoder is enabled) is used (Table 12.5).

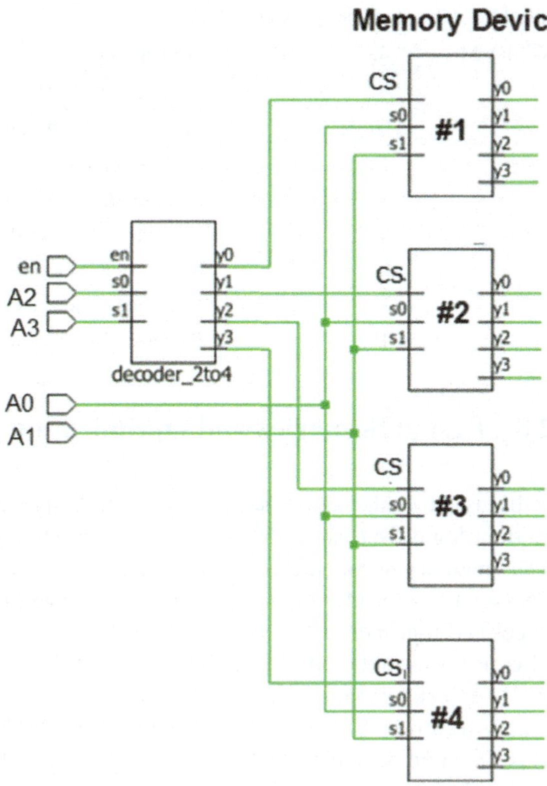

Fig. 12.8 Memory interfacing

Table 12.5 Memory device selection

en = IO_M	*A*3	*A*2	Description
0	0	0	Memory device #1 selected
0	0	1	Memory device #2 selected
0	1	0	Memory device #3 selected
0	1	1	Memory device #4 selected
1	X	X	None of the memory device selected

The address lines *A*1, *A*0 are used directly to identify the memory register or memory location.

The address lines *A*3, *A*2 are used to select one of the memory device at a time using the 2:4 decoder.

Table 12.6 IO and memory selection table

en = IO_M	$A3$	$A2$	Description	Decoder output
0	0	0	Memory device #1 selected	y0
0	0	1	Memory device #2 selected	y1
0	1	0	Memory device #3 selected	y2
0	1	1	Memory device #4 selected	y3
1	0	0	IO device #1 selected	y4
1	0	1	IO device #2 selected	y5
1	1	0	IO device #3 selected	y6
1	1	1	IO device #4 selected	y7

12.8 Design Scenarios and Optimization

As discussed in the above section, we can design the separate IO and memory decoding logic. But using the above discussed strategy, we need to have separate IO and memory decoder. Hence, the system will become bulky and even costlier.

Now let us consider the same scenario to design the system which has four IO devices and four memory devices.

Let us perform the optimization and let us try to implement the single decoding logic to select for one of the IO or memory device. Let us document the IO and memory selection depending on the status of IO_M, RD, WR.

As shown in the table, the IO_M, $A3$, $A2$ are used to identify the selection of the IO or memory device. So let us use the 3:8 decoder and generate the 8 outputs to identify the memory and IO devices.

1. Select lines of 3:8 decoder $s2 =$ IO_M, $s1 = A3$, $s0 = A2$
2. Output lines y0–y7 only one line is active high at a time and according to entries in Table 12.6, used as chip selection lines for the respective memory and IO devices (Fig. 12.9).

Using the above discussion, you can interface the memory and IO devices with the desired processor.

12.9 Concluding Comments

In this manuscript, I have covered the various design elements and their use, design techniques and exercises. We have discussed the

1. Logic gates
2. Boolean functions
3. Combinational elements
4. Design using mux

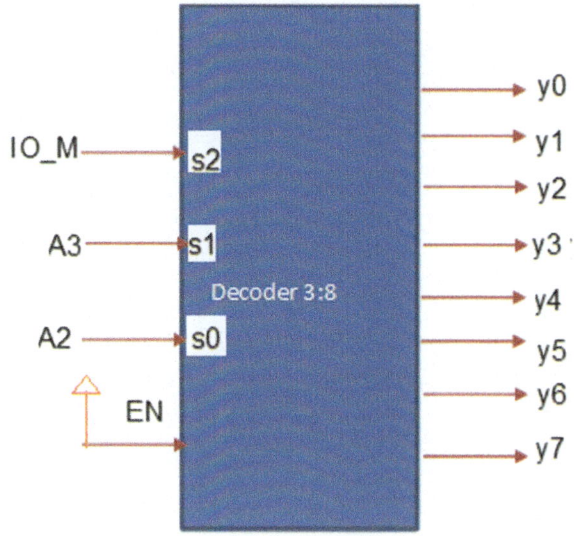

Fig. 12.9 Memory and IO decoder

5. Design using decoders and encoders
6. Flip-flop, latches and applications
7. Counter design
8. Shifter
9. Special counters
10. Sequential design
11. FSM design techniques
12. Area optimization
13. Speed improvement
14. Low-power cells and design
15. Architecture-level design
16. System design
17. Design techniques and optimization concepts.

You can use the foundation and design techniques discussed to learn the advanced design techniques and their role in the VLSI design.

Even you can use the discussed techniques during the

1. Logic design
2. Architecture design
3. RTL design
4. FPGA designs.

Index

V. Taraate, *Digital Design Techniques and Exercises*,
https://doi.org/10.1007/978-981-16-5955-3

The manufacturer's authorised representative in the EU is Springer Nature Customer Service Centre GmbH, Europaplatz 3, 69115 Heidelberg, Germany. If you have any concerns regarding our products, please contact ProductSafety@springernature.com

Printed and bound by CPI Group (UK) Ltd, Croydon, CR0 4YY
30/09/2026
02225291-0002